The Noble Lie

The Noble Lie

When Scientists Give the Right Answers for the Wrong Reasons

Gary Greenberg

WILEY

John Wiley & Sons, Inc.

Published by John Wiley & Sons, Inc., Hoboken, New Jersey
Published simultaneously in Canada

For general information about our other products and services, please contact our
Customer Care Department within the United States at (800) 762–2974, outside the
United States at (317) 572–3993 or fax (317) 572–4002.

Wiley also publishes its books in a variety of electronic formats. Some content that
appears in print may not be available in electronic books. For more information about
Wiley products, visit our web site at www.wiley.com.

Library of Congress Cataloging-in-Publication Data:

Greenberg, Gary, date.
    The noble lie : when scientists give the right answers for the wrong reasons /
Gary Greenberg.
        p.   cm.
    Includes index.
    ISBN 978-0-470-07277-6 (cloth)
        1. Science—Moral and ethical aspects.   2. Science—Social aspects.   I. Title.
Q175.35.G74   2008
174'.95—dc22

                                                                    2008003759

Printed in the United States of America

10   9   8   7   6   5   4   3   2   1

CONTENTS

ACKNOWLEDGMENTS

This book wouldn't exist if not for its editor, Eric Nelson, who noticed the theme of the noble lie in my reporting and then plied me with caffeine until I saw it, too. He's an old-fashioned literary guy—a sensitive reader and a great believer in the power of books, that is—and I am grateful to him.

Many of the chapters that follow began their lives as magazine articles and owe much to their editors. When you stumble into journalism in mid-career, you're even more dependent on editors than you might be otherwise, and I've been fortunate to work with some of the best: Dave Eggers at *McSweeney's,* John Bennet and Amy Davidson at the *New Yorker,* Clara Jeffery at *Mother Jones,* Jennifer Szalai at *Harper's,* and Nick Thompson at *Wired.* I owe them all for encouraging and nurturing me and for saving me from embarrassment on the page. I owe special thanks to Dave and John for taking me on when I was soaking wet behind the ears and teaching me a thing or two.

The stories in this book have all benefited greatly from the careful reading and the thoughtful comments of Bill Musgrave, Jeff Singer, Michael Silverstone, and Rand Cooper. A writer couldn't ask for a smarter, funnier, and sweeter group with whom to take risks. They've egged me on in all the right ways. My breakfast conversations with Glenn Cheney, himself no slouch of a writer, gave many a day's writing a good start.

My production editor, Kimberly Monroe-Hill, ably shepherded the manuscript through the sausage factory. Crackerjack editorial assistant Ellen Wright fielded calls, directed traffic, and provided much needed translation services. I owe special thanks to my mother, Ruth Savin Greenberg, for her sharp-eyed copyediting and for refraining from scolding me for going out on a boat in the rain.

I understand why my psychotherapy clients open up to me: they expect (and sometimes get) something back—support, perhaps, or insight or even something transformative. But I'm still not sure why the people featured in this book let me hang out with them and ask impertinent questions, let alone why they answered them thought-fully and honestly and for the record. I just know that it was mighty generous of them, and I'm grateful for that.

And to Susan and Joel, wife and son, who put up with my absences and preoccupations, my early morning coffee grinding, and my cantankerousness at the end of a long day spent insert-ing and removing commas . . . well, here's a truth that is certain: none of this would have been possible without their indulgence and forbearance and love.

INTRODUCTION

IN THE WINTER OF 1816, René-Théophile-Hyacinthe Laennec, a house physician at a small hospital on the outskirts of Paris, found himself in a delicate position. His patient, a young woman, was complaining of a heart problem, but she was so fat that thumping her chest and listening for changes in resonance, the standard method of diagnostic assessment at the time, was useless for detecting her trouble. The only other method available to him was *immediate auscultation,* the laying of his ear upon her chest with, at most, a silk handkerchief between her bare skin and his. Faced with the equally unpalatable options of ignorance and immodesty, Laennec improvised: he rolled a sheaf of paper from her bedside into a tube, pushed one end through the folds of her flesh until it reached solid ground, and placed his ear on the other. To his great satisfaction, he could now hear her heart clearly.

Laennec, who credited his invention to a couple of schoolboys whom he had seen in the courtyard of the Louvre using a wooden beam to send the sound of a scratching pin from one to the other, never said what became of his patient, perhaps because he had greater ambitions. An amateur woodworker, Laennec soon perfected a wooden version of what he named simply the *cylinder* and began to use it to catalog the sounds of the chest, to which he gave names such as *pectoriloquy* and *rales* and *fremitus*. Thanks to the

1

primitive state of early-nineteenth-century medicine, Laennec was often able to correlate what he heard at the bedside with what he would soon see on the autopsy table. Slowly, the body's inchoate murmurings revealed their meanings, and Laennec was eventually able to use his ear to distinguish pleurisy from emphysema, abscesses from emboli, tubercles from blood clots.

In 1819, Laennec published *De l'auscultation médiate,* a glossary of the thoracic language, and it became a classic text for doctors in training. By the 1830s, other doctors were experimenting with improvements on the device, and in 1851, an American doctor shortened the cylinder and connected it through flexible tubing to a pair of curved metal tubes topped by ivory earpieces, thus allowing him to listen through both ears, not to mention to avoid the indignity of bending over his patients. In this form, Laennec's device has become such a familiar sign of medical authority that a stethoscope slung over the shoulders is a virtual identity card in any hospital and a status symbol in any crowd.

And with good reason. The stethoscope was the first in a long line of devices—X-ray machines, CT scanners, ultrasound detectors—that allow doctors to fulfill the oldest dream of Western medicine: that by using nothing more than their senses (amplified, if necessary) and logic, they can plumb the murky recesses of the body and explain and heal our suffering. Doctors have been pursuing this dream of a purely empirical medicine since Hippocrates (or the group of ancient doctors from Kos who were responsible for the Hippocratic corpus) first insisted that a good doctor must pay exquisitely close attention to the patient,

> to his habits, regimens and pursuits; to his conversation, manners, taciturnity, thoughts, sleep or absence of sleep, and sometimes his dreams; to his picking and scratching; to his tears; to the alvine discharges [i.e., feces], urine, sputa, and vomitings . . . to the sweat, coldness, rigor, cough, sneezing, hiccup, respiration, eructation, flatulence, whether passed silently or with a noise; to hemorrhages and hemorrhoids.

The Hippocratic doctors used their own bodies unsparingly to make this examination, smelling the stools, tasting the urine and the earwax, feeling the skin's temperature and looking at its color, listening carefully to the flatulence. But even this low-tech approach was enough to wrest the understanding of illness from the priests, to kick the gods out of the clinic, and to replace divination and prayer with close observation and reason.

At the heart of this enterprise, which should be familiar to anyone who has ever visited a doctor, is a special kind of knowledge: *diagnosis,* the determination of the truth about our suffering. The ability to examine patients, assess their symptoms, and then reveal what lies behind them is perhaps the most critical skill for good doctoring, the one that makes physicians more than mere technicians of the body. Your doctor knows that your palpitations are *atrial fibrillation* caused by *cardiomyopathy,* that your malaise and thirst are the result of *diabetes,* that your aching back indicates *spinal stenosis,* that even if you're not feeling the least bit bad, you have *hypertension* and *high cholesterol,* maybe even *arteriosclerosis.* Equipped with his special tools, your doctor knows you better than you know yourself.

There is a painting of René Laennec examining a child with his cylinder. A woman, presumably the mother, holds the patient's hand while Laennec turns away, his eyes closed in a concentration that shuts out anything extraneous—the child's chatter, say, or the mother's anxiety, or the doctor's own ambition, or anything that might distort his knowledge, such as human desire or sympathy, political conviction, or religious faith. In his reverie, Laennec appears to be channeling nature, as if his very detachment is the guarantee that the uncertainties of human subjectivity have not contaminated the diagnosis, that the doctor speaks the truth, right from the source.

As men and women of science, doctors don't have an ax to grind. They pronounce their judgments, in the form of diagnoses, solely in the name of health. That's why you listen to them rather than to your brother-in-law or a bartender, why you take their pills, follow their advice, allow them to cut you open and remove the offending part, and, perhaps most important, why you trust them so much that you let their diagnosis become part of your self-understanding. Yesterday you were a person with a cough and fatigue; today you are a cancer patient facing surgery. Yesterday you were unhappy and sleepless; today you are a depressed person starting a regimen of psychoactive drugs that you should follow for the rest of your life, and you submit to this transformation because you believe that your diagnosis is based on an impersonal truth about an indifferent nature rendered accurately by a neutral expert. In that unseen world behind the world, the one to which your doctor has unique access—*nature,* we call it—your disease exists. You are sick because it exists in you, and it doesn't matter what your politics are or how much money you've got or whether you have children whom you love beyond yourself and whose lives would be shattered if your disease kills you. Nor does it matter what the doctor thinks of the disease or of you, or, for that matter, whether he or she even knows anything about you beyond what the stethoscope or the MRI machine divulges.

In a society where pronouncements about what is wrong with us and how we ought to live are always suspect, medicine is where we turn for a truth that cannot be contested, a belief that is not purchased at the expense of fact, a prescription for how to live that is not based in ideology. Science, of which medicine is a crucial part, is the last bastion of certainty based in truth, and doctors and other scientists are the soldiers who hold the fort against blind faith, against irrationality, against unchallenged assumptions. Armed with their scanners and scopes, their statistics and peer reviews and double-blind studies, they patrol the *cordon sanitaire* between the objective and the subjective, between the rational and the irrational, between science and politics, and repel the intelligent designers, the flat-earthers, the New Age flakes, and the snake-oil quacks. And for this hard work, for carving out a place where truth is not a matter of faith but of fact, we pay them the big bucks. Americans spent $2 trillion on health care in 2006, much of it on the bet that doctors know what they are talking about when it comes to suffering and its cure, that when they render a diagnosis, they are doing nothing but faithfully reporting the news from the other end of the stethoscope.

This bet has been institutionalized in many ways. Research grants are generally tied to official diseases, as are insurance payments for office visits, tests, and treatments. The Food and Drug Administration approves drugs only for specific indications, that is, when they prove to be effective for particular diagnoses, and although doctors are free to prescribe drugs off-label, insurance plans will pay only for an indicated use.

And much of the time, the diagnostic wager is a safe one. When a doctor tells you that you have bronchitis and ought to take an antibiotic, the fact that he or she is performing a moral function—telling you what is wrong with your life and how to fix it and, in telling you, making the claim that bronchitis is something we would all be better off without—remains hidden and inconsequential.

But sometimes doctors pronounce diagnoses with deeper significance, in which it matters *why* they think something is a disease. Sometimes, for instance, they tell us that we are unhappy because we have *depression* or that a man has committed serial murder because he has *schizophrenia* or that a person with a devastating brain injury can become an organ donor because she is *dead*. These diagnoses are of great moral consequence. They tell us that we should take antidepressants to alter our consciousness, give a killer psychiatric treatment instead of a lethal injection, or crack open a still-breathing person and take out her heart.

The stories in this book are about some of these diagnoses, the way that they help us to grapple with the unfathomable and set a course through uncharted territory. They are about something that Hippocrates overlooked (or perhaps chose not to address) when he rousted the priests from the clinic: that he was taking on their mantle of authority and passing it along to his successors, some of whom would drape stethoscopes around their necks.

The diagnoses at the center of all these stories have something else in common, something crucial that is often overlooked: they are all invented by people rather than discovered in nature. They are, in other words, fictions. In each story, a diagnosis stands in for philosophy, for religion, for politics—in short, for all the dodgy, tentative ways we have of looking at ourselves—and uses the language of science to settle the question of how we ought to proceed. Which means that these diseases come into existence and survive because we suspend disbelief, because we ignore the evidence—much of it hidden in plain sight, but some of it actively suppressed—that they are made up. The stories in this book are about how badly we need these fictions, how successful they are at guiding us, and how disturbing it is when the consensus they hold together begins to fall apart and we discover that medicine's certainties sometimes come only at the expense of the truth.

One of Hippocrates' contemporaries had a name for this kind of fiction. Writing in *The Republic,* Plato argued that the best way to maintain the stability of society was to claim that its institutions were wrought not by mere citizens but by nature itself. Tell the people that the childhood that they think formed them was an illusion conjured while they grew in the soil like grass or potatoes, and then tell them that they were endowed at birth with varying amounts of precious metals that determined their place in society as the work of *nature;* tell them these things, Socrates said, and you will anchor social order to something transcendent and extrahuman, something that cannot be challenged as the work of fallible humans. The fact that this was not exactly true would be a closely held secret, and the fact that social structure was founded on deception would be justified by the end that it brought about: a just and stable society. Furthermore, the fictions would be as plausible as they were useful. They would be, as Stephen Colbert would have it, *truthy* if not exactly true. Socrates, in Plato's version, insisted that these fictions were noble. The stories that follow are about this kind of fiction, about the noble lies of medicine.

• • •

SOME OF MEDICINE'S FICTIONS are less than noble, purely mischievous even. In 1884, for instance, a letter appeared in the *Philadelphia Medical News* from Egerton Y. Davis, a doctor formerly attached to the U.S. Army. While practicing in England, Davis reported, he was summoned late at night to the manor of a gentleman, whom he "found in a state of great perturbation." The gentleman told the doctor that he had heard a strange noise and, following it to the servants' quarters, discovered his coachman in bed with one of the maids. The pair, upon being discovered, tried frantically to uncouple and finally rolled out of bed, still engaged. The gentleman thought that the maid, who was much smaller than the coachman, was in agony, and sent for Dr. Davis.

> When I arrived, the man was standing up and supporting the woman in his arms, and it was quite evident that his penis was tightly locked in her vagina, and any attempt to dislodge it was accompanied by much pain on the part of both. I applied water, and then ice, but ineffectually, and at last sent for chloroform, a few whiffs of which sent the woman to sleep, relaxed the spasm and released the captive penis.

Davis went on to say that the vaginal muscles had gone into a spasm, which had, in turn, prevented the man's erection from subsiding. Davis speculated that this condition, *penis captivus,* explained a few things: "As an instance of Iago's 'beast with two backs,' the picture was perfect," he wrote, adding that this phenomenon may also shed light on why, in the book of Exodus, Phineas was able to spear both parties to a coupling with one thrust of his javelin.

Over the next century, *penis captivus* cropped up in scholarly papers, mostly from doctors relaying secondhand accounts of the phenomenon, which often occurred in couples having intercourse for the first time. The scholarly papers debated matters such as whether the culprit was the vaginal or the anal sphincter, but all agreed that the best treatment was to chloroform the woman, at least until 1955, when J. S. Oliven proclaimed in his *Sexual Hygiene and Pathology* that the "tried and true" method was "the insertion of a well-lubricated thumb into the woman's rectum." This advice left open the question of whose thumb should be used. And all of these reports referred to the Davis case as the one that established *penis captivus* as an official malady.

It's too bad that none of these doctors consulted the 1925 biography of William Osler, a Johns Hopkins physician and teacher who was famous for insisting that diagnosticians return to the Hippocratic ideal of paying close attention to the data provided by the patient's body. Osler's biographer reported an 1884 conversation in which Minis Hays, the editor of the *Medical News,* asked Osler whether he knew an Egerton Y. Davis, from whom Hays had just received an interesting letter about a hitherto unknown and delicate condition.

"Hays, for Heaven's sake, don't print anything from that man Davis," Osler said. "I know he is not a reputable character."

It turned out, the biography continued, that Osler had an impeccable source for this assessment. He had written the letter himself, having made up the doctor and *penis captivus* in an attempt to parody what he thought was a speculative and pompous paper that had run in the *Medical News* about a kind of vaginismus that was anatomically impossible and revealed more about the doctor's prurient interests than anything else. This was exactly the kind of nonempirical medicine that Osler thought doctors should leave behind.

Egerton Y. Davis continued to bedevil medical journals with articles that included an account of a seal hunter swallowed by a whale and glowing reviews of books by William Osler, but the *penis captivus* caper was the only incident that Osler came to regret. He meant to have a bit of fun, not to sow confusion among his peers and their patients. But he shouldn't have been surprised that it turned out this way, that indeed doctors continued to write seriously about "the Davis case" after the *Medical Times* opined (in 1945) that "as Britons shall never be slaves, so the penis shall never be captured," after scholars in the 1970s formally declared it a hoax, and even after the *British Medical Journal*'s 1979 ban on further correspondence about it (which it temporarily lifted a year later for yet another case study). That was Osler's point: that the power of diagnosis was self-perpetuating, that as physicians more and more claimed science as the source of their proclamations, they also gained the power to name our suffering and thus to make their sober pronouncements about even the most intimate parts of our lives. If you dress it up in the trappings of science, Osler was saying—and it doesn't hurt if the case has a bit of an erotic frisson—even the most far-fetched notion can be made to seem plausible, and any human foible or frailty can be turned into a disease, ready to be diagnosed and treated.

Osler's broadside misfired, most likely because he had committed the cardinal error of the amateur satirist: overestimating his audience's critical distance from their sacred cow. But in the 125 years since the Davis case, doctors have frequently demonstrated the validity of his complaint. Especially the doctors who work for drug companies, like John Winkelman, a Harvard doctor, who in 2003 warned Americans about a "common yet under recognized disorder . . . keeping America awake at night": *restless legs syndrome* (RLS), "an uncontrollable urge to move [the] legs, or 'creepy-crawly' sensations in the legs . . . that often leads to sleep disruption." Winkelman, speaking on behalf of GlaxoSmithKline (GSK), informed us that RLS "can produce severe insomnia and difficulties with daytime functioning." At the same time, the National Sleep Foundation reported a study (also funded by GSK) showing that while "17 percent of adults 55 to 84 reported unpleasant tingling feelings in the legs . . . only five to seven percent said they had been diagnosed." This, according to Dr. Winkelman, was a shame. "Individuals with RLS should not suffer but instead talk with their physician about a treatment plan." Which is where GSK came in. The company just happened to have a drug that relieved RLS—Requip, a treatment for Parkinson's disease that had been only a middling performer.

The *Movement Disorders* article showing Requip to be an effective treatment for RLS didn't come out until the end of 2004, and the Food and Drug Administration didn't give GSK the approval that allowed it to advertise that fact until the following year. But that didn't mean that the announcement (or the presentations preceding it at the American Academy of Neurology's 2003 meeting about Requip and RLS) was premature. In fact, those two years before Requip's official launch were GSK's opportunity to convince millions of Americans that they had the disease for which the drug was the cure. And the company had the perfect outlet for this campaign—not advertising, which, of course, most people reflexively view with skepticism, but the news media. The discovery of an under-reported illness, one that silently afflicts people, ruins their sleep, and even drives them out of their spouses' beds—this is *news*. Health and science reporters wrote as many articles about RLS between 2003 and 2005 as they had in the previous decade. By the time the Food and Drug Administration made its decision about Requip, RLS was an official diagnosis, not only in the medical books but in the minds of medical consumers. This impression was only strengthened when, in 2007, scientists found a genetic variation that was present in many people who complained of RLS, a bit of news that the *New York Times* suggested might lead to "more respect" for its sufferers. It was as if the fact that a condition had some biological correlate *proved* that it was really a disease.

Erectile dysfunction, premenstrual dysphoric disorder, excessive daytime sleepiness, restless legs syndrome: these and other diseases are now familiar enough to be fodder for Jay Leno jokes and even a Rush Limbaugh parody. The process of launching a diagnosis in order to launch its cure even has a name—*disease mongering*—bestowed by the same media that monger the diseases. It is a predictable, if perverse, outcome of a health-care system that bestows enormous rewards on drug makers, doctors, hospitals, and universities, but only for treating specific diseases. Disease mongering is, in this sense, the bastard offspring of the marriage between science and the free market, which we trust to be almost as neutral as science in its determinations. Science names our maladies as diseases, and the free market provides the cures.

With $2 trillion at stake, however, the naming of diseases is, from a shareholder perspective, too important to be left entirely in the hands of the scientists. But even if a diagnosis is a potential ticket to riches, it would be a mistake to think that disease mongering is the sole province of the pharmaceutical industry or that it's always done for profit. Neither is it a new phenomenon, something that emerges only in our media-saturated, bottom line–driven times. Indeed, some of our most venerable diseases are the handiwork of public relations experts.

BEFORE THE *Quarterly Journal of Studies on Alcohol* ran his article "Alcohol and Public Opinion," Dwight Anderson had never written for a scientific journal. He was a marketing man, the chairman of the board of the National Association of Publicity Directors. Unlikely as it seems, however, Anderson's article played a seminal role in creating the disease we now call *substance dependence*. It was 1942, Prohibition was less than a decade gone, and doctors everywhere were dismayed by the failure of their profession to find effective treatments for what at the time was still called *inebriety*. They were even more dismayed at their difficulty in attracting public attention to the need for medical research into the problem. It was as if Prohibition had frozen the public perception of chronic drunkenness in its nineteenth-century form, as a problem for ministers and cops, who would call in the doctors only to supervise the drying-out or to attend to complications like cirrhosis. Some of these doctors formed the Research Council on Problems of Alcohol to figure out how to get recognition as the go-to guys for alcohol problems, and they turned to Anderson for some advice. A recovered alcoholic, he rendered a quick diagnosis: the doctors had failed to make sufficient hay of the idea that chronic drunkenness was actually a physical disease. In his article, Anderson laid out the case for the diagnosis:

> What are the ideas of the least common denominator concerning alcohol which can be most easily established . . . ? The first is that the "alcoholic" is a sick man who is exceptionally reactive to alcohol. . . . Sickness implies the possibility of treatment. It also implies that, to some extent at least, the individual is not responsible for his condition. It further implies that it is worth while to try to help the sick one. Lastly, it follows from all this that the problem is a responsibility of the medical profession, of the constituted health authorities, and of the public in general. . . . When these ideas have been fully accepted by a large number of people . . . the "yes" response becomes automatic, uncritical, and on the emotional level . . . and only by this means can the required approvals be gained for changing existing situations, for the creation of new institutions, for the formation of groups to *do things* without which science remains inert.

Anderson proposed using the principles of modern marketing to solve the problem of chronic drunkenness: convince consumers that their suffering is related to a deficiency that only the client's product can relieve. To judge from the subsequent success of the disease model of alcoholism, this was brilliant advice.

Anderson was hired to write his article by Elwin Morton Jellinek, a statistician who was preparing a monograph for the Research Council that it hoped would demonstrate the grave threat that inebriety posed to the public health. It's no surprise that Jellinek turned to someone who had never before written a scientific treatise. His own life had prepared him to understand the limits of facts, to know when invention was necessary. Born in New York and raised in Budapest, where he acquired the nickname "Bunky" (Hungarian for "little radish"), Jellinek was nearly fifty years old when he joined the nascent Research Council in 1939, and his only experience with alcohol to that point was whatever drinking he had done (and he never said how much that was) and tending to friends during their binges. He did have some academic experience—studies in liberal arts at Leipzig and Grenoble—and spoke or read twelve languages, but he never really held the master's and doctoral degrees that he claimed Leipzig had awarded him, he never fully explained why he had disappeared from Budapest (where, according to his daughter, he did something related to foreign currency) and surfaced in Sierra Leone as a businessman with the name Nikita Hartmann, and he couldn't quite account for his sudden move from there to Honduras or how he convinced United Fruit that he could research plant biology for them. But when he left Central America, he landed a job—perhaps on the strength of his doctorate from the University of Tegucigalpa, which he had added to his résumé and which also turned out to be bogus—at Worcester State Hospital as a statistician and an editor of a professional journal. When he went to work for the Research Council, it was explicitly for his editing skills, but it was his capacity to navigate the political and social landscape of drinking that ultimately served the organization (and Jellinek) best.

Six years after Prohibition was repealed, the tortured politics of wet and dry were still making coherent social policies about alcohol nearly impossible. Americans could not agree on just what kind of problem inebriety was, and they never had. Drinking had featured prominently in American life since the earliest colonists had unloaded casks of rum and ale and wine in Boston Harbor. Puritans drank at church celebrations and barn raisings and military drills, in court and legislative sessions, and in the taverns where they hatched the seditions of the rebel nation. Even stern Cotton Mather called alcohol "the good creature of God." But at the same time, no one could ignore the many drunkards among the colonists, and by the end of the eighteenth century, the preachers of the Great Awakening had come to see alcohol not only as a temptation but as the litmus test of a soul's destiny. "When a drunkard has his liquor before him," Jonathan Edwards thundered from the pulpit, "he has to choose whether to drink or no. . . . If he wills to drink, then drinking is the proper object of the act of his Will"—a will that Edwards thought issued from a corrupt soul. If a drunkard couldn't help himself, that was only an indication that he was one of those sinners whom an angry God held in His hands, that he was thus in dire need of redemption.

But another prominent early American, Benjamin Rush— a signer of the Declaration of Independence, a doctor to the Continental Army, and a proponent of a constitutional guarantee of the right of medical freedom (lest health care become the entitlement of the rich)—announced in 1810 that "habitual drunkenness should be regarded not as a bad habit but as a disease . . . a palsy of the will." With Edwards, Rush believed that abstinence was the best thing, but not because sinners should be prevented from sinning. The problem, Rush thought, was not spiritual but physical, not moral but medical, not a matter of right and wrong but of health and sickness. Alcohol, he argued, infected the soul, incapacitating the free will that all men in good health naturally possessed. Like many patriots of his time, Rush believed that the supremacy of democracy itself hinged on the belief that God intended us to live in freedom, that He endowed us with the capacity to do so. "Palsy of the will" was thus not a political problem, not something to argue about, but an indication that something had gone wrong in God's machinery, in the body that was now the province of doctors. It was also a public health problem and an urgent one at that—so grave, Rush thought, that the pathogen should be immediately banished. A century later, temperance groups, which claimed Rush as their founder, succeeded in fulfilling his dream of Prohibition.

The spectacular failure of Prohibition, thanks largely to the way it criminalized a large sector of the public and led to bathtub gin scourges like "jake leg" and blindness, not to mention the violence of bootlegging, had forced mandatory abstinence into disrepute. The doctors at the Research Council, however, believed that abstinence was the only answer for the inebriate, and since 1935, they had had at least two allies. Bill Wilson, a stockbroker from New York, and Bob Smith, an Akron, Ohio, doctor—the first recently dried out, the second in the process of drinking his practice into the ground—were introduced to each other by the minister Wilson had called from his Akron hotel room where he had found himself craving a drink. Smith and Wilson discovered that even though they were strangers, their mutual indenture to alcohol linked them intimately. They talked each other out of drinking and decided to start a new organization that would be a fellowship of drunkards who would help one another stay sober one day at a time.

Alcoholics Anonymous (AA) started small, with chapters in New York and Akron and soon in Cleveland, but Wilson and Smith had larger ambitions. They also faced major obstacles to getting out the word that there was hope for chronic drunks. Not only did they favor abstinence, but they were an offshoot of the Oxford Group, an evangelical Protestant organization (which eventually changed its name to Moral Re-Armament) that advocated "Four Absolutes": honesty, purity, selflessness, and unbounded love. Its confessional prayer groups were the model for the AA meeting. An ascetic Christianity that favored abstinence was too reminiscent of Carrie Nation for a country so recently traumatized by Prohibition. By 1939, the Cleveland branch had closed, and Wilson's book, *Alcoholics Anonymous: The Story of How More Than One Hundred Men and Women Have Recovered from Alcoholism,* was a commercial dud.

But then AA got some lucky breaks. First, there were some high-profile successes, including Rollie Hemsley, the Cleveland Indians player most famous for being Bob Feller's catcher, who confessed his drunkenness in the national press and credited AA with his reform. But even more important was what happened when Sally Mann, a journalist who claimed to be the first woman who achieved sobriety through AA and who made it her life's work to spread the AA gospel, met Bunky Jellinek.

By then, Jellinek's work at the Research Council had earned him the ultimate ticket to respectability: an appointment to Yale, where he joined the staff of the Laboratory of Applied Physiology in 1941 and became the managing editor of the *Quarterly Journal*

of Studies on Alcohol, two institutions that led the way in research
and treatment of the new disease. And Mann was ready to spread
the word about just what kind of disease inebriety was. It was
all spelled out in a chapter of *Alcoholics Anonymous:* "A Doctor's
Opinion," written by William Silkworth, the physician who
ran the upper-crust drunk tank where Wilson had detoxed. In
it, Silkworth brought Rush's ideas into the twentieth century,
opining that inebriety was *alcoholism,* a word that doctors had
previously used to refer to the effects of chronic drunkenness, and
drunkards were *alcoholics,* who had an "allergy" to alcohol
and thus "cannot use liquor at all, for physiological reasons." The
fault, Silkworth said, was not in the bottle but in ourselves, at least
in those selves unlucky enough to inhabit sick bodies. Alcoholism
is the *cause,* rather than the *effect,* of inebriety, something over
which the will had no control because it originated elsewhere;
chronic drinking was the outcome of "a law of nature operating inexorably." Alcoholism, in other words, was a disease in the
most modern sense of the word, and modern man—that is to say,
rational man—will "accept the situation . . . and shape his policy
accordingly" by abstaining from alcohol. And a rational society,
Silkworth thought, will put alcoholics in the hands of the men
with the stethoscopes and give doctors the resources to find the
best way to help patients achieve sobriety.

Silkworth's disease and Mann's tireless advocacy of it were
exactly what the Research Council had ordered—a way to avoid
the wet-dry culture wars while still getting out the message about
chronic drunkenness. By taking alcoholism entirely out of the
moral realm and into the medical, the allergy model at a single
stroke offered reassurance to all interested parties: alcoholics
would get treatment in place of moral condemnation, drys could
maintain that there was still something wrong with drinking
(albeit only for some people), wets could argue that there was a
place for alcohol in American life, clergy could still exhort (some)
people toward (physician-assisted) abstinence and open their
church basements to AA groups, and the newly reinvigorated

brewing and distilling industries, which were pleased to help fund the Research Council's research, could claim that science had proved that alcohol didn't kill people; *alcoholism* did.

Jellinek's section of the laboratory eventually became the Yale Center for Studies of Alcohol, which in turn started the Yale Clinics and sponsored classes in alcohol studies at the Yale Summer School. In 1944, Jellinek helped Mann to start the National Council for Education on Alcoholism in order to inform Americans of "two momentous discoveries":

> FIRST, that alcoholism is a *sickness,* not a moral delinquency. SECOND, that when this is properly recognized *the hitherto hopeless alcoholic can be completely rehabilitated.*

Jellinek threw his Ivy League weight behind these ideas, Mann campaigned tirelessly on their behalf, and Alcoholics Anonymous gathered members in their wake. By 1949, twelve states were sponsoring alcohol treatment programs, all of them run by graduates of Jellinek's Yale Summer School and operating on the belief that alcoholism was a chronic disease for which AA attendance and lifelong abstinence were the treatment. Many more programs would follow. The "yes" response, as Anderson had predicted, was becoming automatic.

BUT THERE WAS ONE CATCH. The disease model may have been brilliant public relations, but it was not very good science, at least not yet. In what sense is alcoholism a disease? What exactly is wrong with the people who can't control their drinking (other than the fact that they can't control their drinking)? What is the mechanism of this "allergy"? Jellinek struggled with these questions and found himself up against the triumphs of modern medicine. Smallpox, syphilis, diabetes: these and other signal discoveries had raised the bar for establishing a valid diagnosis. In each case, a single pathology—a virus, a spirochete, a malfunctioning pancreas—proved to be the underlying cause of illness and led to a cure: inoculation, antibiotics, insulin injections. To call something a disease was to imply that something *physical* would be discerned with a stethoscope or its modern equivalent and found responsible. But the proponents of the allergy model abandoned it in 1952, and the pioneer neuroscientists of that decade failed to find the kind of biochemical indicators for alcoholism that they were finding for depression and schizophrenia. Nor did endocrinological or nutritional or cellular studies discern the kind of differences between alcoholics and the rest of the population that could definitively be called causes, rather than effects, of drinking. The disease model was turning out to be too good to be true.

By 1960, Jellinek had to acknowledge the difficulty when he spelled out his theory in *The Disease Concept of Alcoholism*. He defended the flaws in the concept by arguing that it was a mistake to insist on finding an underlying pathology before accepting that a particular condition was a disease. "The fact that [doctors] are not able to explain the nature of a condition does not constitute proof that it is not an illness," he wrote. "There are many instances in the history of medicine of diseases whose nature was unknown for many years." Absence of evidence, Jellinek claimed, was not evidence of absence, and not because future findings might finally turn up the pathogen, but because disease itself wasn't really something in nature after all. "It comes to this," he announced in italics, "*a disease is what the medical profession recognizes as such.*" It seemed that the physician's authority to say which forms of suffering were diseases could be cut free from the science that justified it.

Dwight Anderson or, for that matter, Egerton Y. Davis himself couldn't have said it better, although perhaps neither would have dared to say it so baldly. And both would have been impressed with the continued success of Jellinek's disease, much of which Jellinek himself, who died in 1962, didn't get to see. In 1965, the American Psychiatric Association voted to admit *alcoholism* to its nomenclature, and a year later, the American Medical Association followed suit. The National Institute on Alcohol Abuse and Alcoholism was founded in 1970. By 1973, all fifty states had programs to treat alcoholism, as did most hospitals, and insurance companies paid for inpatient treatment, nearly all of which was based on the allergy-abstinence model. Research dollars flowed, as scientists increasingly focused on what they saw as the disease mechanism that was central not only to alcoholism but to other drug problems, and eventually to behaviors like gambling and sex: *addiction,* the disabling of the will. And Americans increasingly came to see their compulsions, their difficulties in moderating not only their drug consumption but their eating, gambling, shopping, having sex, working—indeed nearly any activity—as the symptoms of this new illness.

But while scientists have unveiled some of the genetic and neurochemical correlates of addiction and developed some drugs that help to block cravings and other withdrawal symptoms, no one has yet discovered the pathogen, the "law of nature operating inexorably," that Silkworth thought lay behind addiction. *Alcoholism, addiction, substance dependence*—these terms perhaps sound more scientific than "inebriety" or "palsy of the will," and present-day doctors can talk knowledgeably about dopamine metabolism and other mechanisms that Benjamin Rush may only have dreamed of, but the idea that there is something in the body that afflicts certain people and renders them incapable of exercising their free will over alcohol remains just that—an idea, a fiction.

That hasn't stopped the disease model from becoming the conventional wisdom about addiction, the reason that most of the $5.5 billion a year we spend on treating addiction goes to doctors and hospitals. Perhaps this is because it's such a good idea, and in a way that neither Jellinek nor Anderson knew. The idea that addiction is an illness for which sobriety is the cure helps us to negotiate some of the vast confusions that have always haunted American life: our ambivalence about pleasure (especially drug-induced pleasure), for instance, or the uncertainties about the limits of free will and self-determination, a culture thrash that started before

Jonathan Edwards and Benjamin Rush and continues today. These questions threaten to emerge whenever we see a person in the throes of addiction; but with the disease model, we have a ready-made answer, one that has the imprimatur of science: addiction isn't *wrong,* it's *sick;* abstinence isn't *virtuous,* it's merely *healthy,* and then only for those with the affliction. And when you tell a person that he is drinking too much, you aren't exercising a moral judgment. You're simply telling him that he has a disease.

There can be no doubt that the disease model has helped millions of people. If a made-up disease can be of such immense value, then we must consider the possibility that the truth is not all it's cracked up to be. Perhaps in the republic of medicine, the fiction that addiction is a disease is a noble lie.

MOST HISTORIANS and social commentators object to noble lies on the grounds that in a society in which we are free (and expected) to figure out our own lives, we must all have equal access to the truth, that an elite that alone knows the truth is bound to become a cabal or an oligarchy. (Leo Strauss, the godfather of the neoconservative movement in the United States, was a notable exception; he thought it was an excellent idea for an executive group to advance the foundational fictions that would render irrelevant the reality-based community.) But when it comes to medicine, to keeping ourselves alive and well, perhaps we needn't worry. Surely, addicts are better off in hospitals tended to by doctors in the service of health than in public stocks at the urging of hellfire-breathing ministers who serve salvation. Surely, fiction in defense of health (and at the expense of intellectual liberty) is no vice.

On the other hand, consider Howard Lotsof, whose quest to change the way we think of and treat addiction you'll encounter later in this book. Forty-five years ago, while a junkie in New York looking for a new kick, Lotsof took a preparation of the powdered root of iboga, a hallucinogenic plant from Africa. He came out of the twenty-four-hour-long experience suddenly freed from his addiction, and he never returned to heroin. He did this without thinking of himself as someone with a chronic illness that has to be battled one day at a time and that requires perpetual sobriety and attendance at meetings. And he's been trying to get governments, industry, scientists—anyone who will listen, really— to develop ibogaine as an antiaddiction drug. But his one-shot treatment undermines the pharmaceuticals' one-a-day business model and the disease model to which it is wedded. Even mainstream academic researchers find it difficult to obtain funding for a treatment that doesn't start with the assumption that addiction is a disease that never really goes away. Indeed, the people who have come closest to success in developing ibogaine are those who are trying to turn it into a patentable medicine that can be taken every day.

But it would be a mistake to chalk up the persistence of this fiction to drug company greed. The disease model of addiction fits American society like a glove on a hand. It helps all of us, addicts or not, to understand ourselves in ways that make living here easier; we don't have to fight about whether the addict is a sinner or what it means that free will can be subverted so easily. The noble lies in the stories that follow all share this quality with Plato's original deceptions. They help us not only to figure out who we are and who we ought to be but also to know why, with certainty—because, scientifically speaking, that is the way things are.

And look what happens when we don't have medicine's noble lies to guide us. One of the most vexing features of modern cultural life is the interminable wrangling over questions raised by biotechnology—the paralyzing debates about stem-cell research, abortion, and assisted suicide, and about steroids and "smart drugs" and other enhancement technologies. We haven't yet agreed on fictions about these matters, or, more accurately, when it comes to these technologies, there is no noble lie that allows us to get on with the business of medicine, and there may never be. Doctors don't have the authority they once did; pharmaceutical companies have recently taken huge, well-deserved hits to their credibility; bioethicists are hamstrung by their nearly total focus on the process by which we reach decisions about biotechnology rather than on their content; and presidential efforts to mandate such conclusions are easy to unmask as attempts to speak power to truth. As biotechnologies come to focus more and more directly on changing what it means to be human, the confusion only deepens, and the disappearance of noble lies signals a crisis—in science, in medicine, in our self-understanding.

The stories in this book will not tell us how to resolve this crisis. But they will make it clear that science is not going to save us from them and why this is so: because the answers to these questions are not to be found in nature, no matter how carefully we look, no matter how objective we try to be. The line between illness and health will always change with new knowledge and improved technology and even with shifting fashion, and not only when it comes to the diagnoses found in this book. Because Bunky Jellinek had it right: diseases are what the medical profession says they are.

There are better diseases and worse diseases; fictions about addictions are arguably more valuable than fictions about restless legs. But diagnoses will always be fashioned according to prevailing notions of the good life and the good person, of what kind of people we ought to be. To give suffering a scientific name is not to remove it from the hurly-burly of human history, much as we might wish. Indeed, as each of the following stories illustrates, there are no diseases in nature. The activist and philosopher Peter Sedgwick has put this beautifully:

> The fracture of a septuagenarian's femur has, within the world of nature, no more significance than the snapping of an autumn leaf from its twig; and the invasion of a human organism by cholera carries with it no more the stamp of "illness" than does the souring of milk by other forms of bacteria.

There's plenty of suffering in the human world, but none of it matters until we give it a name. Once we've done that, we can put our doctors and the vast apparatus at their command to work to relieve it. To have a disease is to have a claim on those resources, which, enormous as they are, are still limited. The stories that follow make it clear that the work of deciding which suffering should be relieved, and how, is not as simple as placing a stethoscope to a chest and listening to what nature has to say.

ADDICTION:
VISIONS OF HEALING

F OR A GUY WHO'S JUST GOTTEN into a car with total strangers and let them whisk him onto a boat, Moob—it's short for something, but he won't let me say what—seems pretty calm. He's on the prow of one of the big ferries that ply the archipelago just north of Vancouver, the wind mussing his thick, dark hair, sea mist collecting on the brown leather jacket that he says he got off a dead junkie. It's not clear which came first—the death or the jacket—but Moob definitely seems like the kind of guy who could have helped himself and beat feet just before the cops showed up. Not that he doesn't seem decent enough right now, smooth and relaxed and even a little charming, as we chat about his childhood fireworks obsession, about forest fires, and about the effects of cocaine on the sphincter. But I've met plenty of addicts, and Moob has that not-quite-dialed-in demeanor, like he's watching our little drama unfold from a perch on the moon—the same detachment that has probably allowed him to put himself in this situation in the first place—and I'm sure that if any of us dropped dead on the deck, he wouldn't hesitate to relieve our corpses of their earthly burdens.

From what I can tell, he wouldn't get much off of Linnette Carriere, the pretty twenty-five-year-old woman whose long sandy braid descends from a jaunty beret. She's wearing a dowdy wool pullover, she drifted to British Columbia from a murky past in the eastern provinces, and she seems anything but afflu-ent. And I didn't realize that I was embarking on a two-day jour-ney when I was summoned a little while ago from my cozy hotel into the raw and rainy British Columbia winter, so I don't even have a toothbrush on me. But the other man with us, a handsome and fit forty-five-year-old, has his own leather jacket—a step up from Moob's, creamy and gorgeous. He lives in a downtown luxury high-rise with a view of the Strait of Georgia, where he conducts his multimillion-dollar business and his complicated love life (which no longer includes Carriere). He drove us onto the ferry in a shiny new car. And he's no doubt carrying a big wad of Canadian money, because when he isn't conducting people to the Iboga Therapy House, his private rehab facility on British Columbia's Sunshine Coast, Marc Emery is Canada's Prince of Pot, the self-proclaimed largest purveyor of marijuana seeds in the world—a business that he conducts entirely in cash.

But if anyone is going to die prematurely here, it's Moob, who is thirty-six and olive-skinned, a construction worker sporting a mustache that draws attention to the gap where his right front tooth ought to be. And not only because he usually associates with junkies or because he has been known to smoke a couple thousand dollars' worth of crack in a binge, so much that even his crackhead friends have told him it's time to clean up. It's also because after we arrive on the Sunshine Coast and drive to the house, as Carriere helps and I watch, Emery is going to give Moob a few grams of the powdered root of *Tabernanthe iboga,* a bush that grows in West Africa, where it is used as a ritual hallucinogen. The drug induces a long, arduous trip, twelve or twenty-four or even thirty-six hours of nausea and dizziness, featuring a kinescopic life review that is heavy on scenes of moral failure, a searing journey sometimes led by a hallucinated spirit guide who wields a large stick. It also causes your pulse and breathing to slow way down, an effect that in combination with imprecise dosing or the residue of street drugs or just unlucky genetics sometimes proves lethal. This fact is on my mind, if not on Moob's, when Emery announces, to no one in particular, mostly to the wind and the rain, "Of course, I *am* practicing medicine without a license here."

Moob does seem to be aware of some risks or at least of the overall strangeness of this trip. "I've heard some stories," he says, and so have I—about how addicts come back from their ibogaine journeys without their back-monkeys, somehow miraculously propelled through withdrawal and beyond craving and into a world where the drug they were hooked on holds no interest. I heard it back in Vancouver from Sheldon, a twenty-five-year-old who had been on the city streets for nearly ten years and shooting and smoking heroin for six. He was the first addict to take Emery up on his offer to provide ibogaine treatment free of charge to anyone who wanted it. (Emery learned of ibogaine when an employee had returned from a treatment clean and sober.)

"I didn't know anything else about it," Sheldon told me over coffee in Vancouver. "I just knew that every detox was filled and even the psych wards wouldn't take me." But Emery, who is a local hero for his willingness to take on the drug war (he is a perpetual mayoral candidate on the BC Pot Party's "Overgrow the Government" campaign and a major source of funds for the legalization movement), inspired enough confidence for Sheldon to give it a try.

"It was hours and hours of visualizations that were personal and truthful and really, really hard," Sheldon said, as we sat at an outdoor café in the rain. "Stuff I would never think about came to me. And I saw myself. I saw how selfish I'd been, how I affect other people. Like someone was saying to me, 'You're twenty-five years old; you got to grow up.' Which is stupid, of course. I should have already known that. But I didn't. Just like I didn't see, till the ibogaine, that you just don't know how long you're gonna live and you have to deal with that. You have to account for yourself." Sheldon came out of his trip without dopesickness but, more important, with a new view of his life. "It just didn't make sense to use anymore." The visions had fashioned a new moral universe, one in which what did make sense was to enroll in film school, get in touch with his long-estranged family, and try to get his younger sister off the streets and into ibogaine treatment. "I still get the cravings, but I listen to them and watch them and I don't have to act on them," he said. "It's an easy decision."

This is the outcome Moob is betting on. Like Sheldon, he is a veteran of hospital treatments that didn't work, of twelve-step groups that he didn't fit into, and of cold turkey resolve that has gone up in the first available smoke, so it's a simple calculation: Moob has the jones and Emery has the cure waiting for him at the other end of this ferry ride. If this goes against everything Moob has heard before about addiction—that it's a chronic illness, that there is no cure, that recovery is achieved one day at a time and only after you've surrendered to the Higher Power and begun to work the steps—it really doesn't matter, because that approach hasn't worked for him. "If it's going to get me off the crack, then anything is worth a try," he says.

But for other people—addiction doctors, drug makers, government regulators—ibogaine is a problem. Since the success of Dwight Anderson's plan to give doctors control of addiction treatment, addiction has been turned from a sin in need of redemption into a disease necessitating a lifetime of recovery. The entire edifice of addiction treatment is raised on this conceptual infrastructure, on the idea that addiction is something to be dealt with by the detox doctors and the AA groups and, in some hoped-for future, by the drug inventors who can find a way to control this chronic disease. There is no doubt that many have benefited from this approach, but if a single frank and bracing look at yourself, presided over by a pot seed salesman and his vagabond ex-lover, can do what Moob is hoping it does for him, then that model, and the doctors and therapists and hospitals and pharmaceutical companies it supports—not to mention our common beliefs about drug use and sobriety—could be in trouble.

MARC EMERY ONCE BOUGHT a table for himself and his friends at a Vancouver fund-raiser whose featured speaker was John P. Walters, then the U.S. drug czar, who was in town to warn the Canadians not to decriminalize marijuana. The drug was probably never the harmless wacky weed that many people once thought it was, he argued, but these days it was so much more potent, especially the stuff grown right here in British Columbia, that if the Canadians decided to stop protecting their kids, the United States might have to close its borders for the safety of its own.

After tricking the czar into posing for a picture with him, Emery and his friends heckled him from their table. (The drug czar may get the last laugh. In 2005, the U.S. Department of Justice indicted Emery as a drug kingpin, charges that carried the possibility of life imprisonment. Emery fought extradition for nearly three years, claiming that he had already done his time and paid his fine in Canada, whose government had then decided to tolerate his operation. In early 2008, however, Emery announced that he'd reached a tentative deal that would allow him to serve five years in a Canadian prison. In the meantime, his seed business has been shut down.)

Provocations like these are Emery's specialty, but it's what he did next that makes him an unlikely guy to run a rehab. After the luncheon, he told me, he went outside, lit up a big joint, and passed it around. That probably wasn't the first time he'd smoked that day. Emery smokes huge quantities of the green menace and thinks that we all, addicts and otherwise, should be free to do so. And this violates a central belief of the addiction treatment industry. William Silkwood's idea that addiction is an allergy has morphed into the belief that if you're allergic to one drug, then you must be allergic to all of them, as if an allergy to bee stings meant you would also be allergic to nuts. This isn't just an abstract idea: some clinics that refer patients to me prohibit even aspirin during the rehab period.

Of course, the ban on drugs doesn't include psychiatric drugs, although some doctors are wary of prescribing Valium or Xanax even to people who have never abused them. But these exceptions aside, the prevailing belief of recovery is that if you are taking a drug to change the way you feel, then you are enacting your disease. That is, sobriety is not simply a matter of not taking the substance that you got hooked on but of being clean and sober. This is curious, not only because your average AA meeting is a caffeine-soaked, nicotine-fogged affair, but because the ban on being high does not extend to "natural" highs, like those achieved by long-distance runners. The condition you have to recover from, in other words, is a compulsive need to turn to something outside yourself to make yourself feel better.

This *dependence* is an affront to a society that values independence as highly as ours does. Addiction once carried a different meaning. The word derives from the Latin *ad* and *dictum*—literally, "toward the dictum," or "obedience"—and it used to describe the relationship between an apprentice and his master. An addict turned his life over to the force outside himself, and in the days when apprenticeship was the way to learn things, and when people weren't quite as concerned about autonomy as we are, it wasn't pathological to be addicted in this fashion. Of course, drug addiction—at least, its physical ravages—would be a problem even in a feudal society. But it would be a different kind of problem if it weren't informed by this suspicion of dependence, if there weren't already a belief that hard work and self-sufficiency were the only legitimate ways to pursue happiness.

Consider the case of steroids in sports. Why is it considered cheating to use them, other than the fact that they are banned, which only invites the question of why they were banned in the first place? After all, sports performance in general has improved over time, at least to judge from the fact that records keep toppling. Runners run faster, jumpers jump higher, sluggers slug harder. Some of this is due to more rigorous (and scientific) training programs—weight training, agility training, endurance training, and even psychological training—all designed to help athletes make the most of their gifts. Some of it is due to the improved nutrition, sanitation, and medicine that have made most of us taller and stronger and longer-lived than we would have been a century ago. But neither advanced training regimens nor partaking of progress in the public health sphere have been banned from sports.

This is not because these are entirely harmless ways to better ourselves. Athletic training can cause injury and certainly alters the body you were born with, sometimes in less than optimal ways; think of how a tennis pro's playing arm is noticeably larger than his other arm or how women athletes' menstrual cycles often stop when they are in top shape. Nor is it because they are "natural" methods; on the contrary, they often rely on sophisticated machinery and scientific research and sometimes require exercises that seem quite unnatural. But imagine for a moment what would happen if a food was developed that, when eaten, did what steroids do: increase muscle mass by ramping up protein synthesis and the amount of testosterone (which decreases the transformation of muscle into fat) in the body. This is not a far-fetched idea when you consider that many steroids start their lives as synthetic or natural versions of animal hormones, especially testosterone. (The first reported use of steroids was by a seventy-two-year-old British doctor who injected himself with an extract made from dog and guinea pig testicles, which he said made him feel rejuvenated.) Although such an invention might not pass muster with people who object to genetically altered foods, its availability would very likely change the debate over steroids because steroids would no longer be taken in needles and pills and creams, but instead by eating a steak from a particularly virile animal. Steroids would, that is, be "food," which is something we *must* ingest, something we can take in from outside ourselves without worries about dependency (unless you're a "food addict," which is another story).

If steroids were no longer drugs, they might no longer be considered cheating, because they wouldn't violate the rule against employing outside agents in your pursuit of happiness. Self-reliantly toiling by the sweat of our brows to achieve sanctioned pleasures is a cardinal virtue of Western civilization, a point spelled out more than a hundred years ago by Max Weber in *The Protestant Ethic and the Spirit of Capitalism.* The need for this discipline, as Weber showed, derives from a singular notion about human nature—that we are dissolute and lazy creatures who must whip ourselves into shape lest we succumb to "the temptations of the flesh." This fiction (for all accounts about human nature are necessarily stories) is what gives purchase to the idea of "sobriety" on which the drug treatment industry is built. We are already deeply suspicious of powerful agents outside ourselves, especially those that deliver pleasure, and the addict stands for what happens when we surrender our autonomy in the pursuit of happiness. The allergy isn't to a particular chemical but to all chemicals that induce the loss of self.

The fiction of the autonomous self has made us into the hard workers that we are. The fiction is noble in this respect, for, as Weber pointed out, it is directly or indirectly responsible for the breathtaking achievements of modernity. But it doesn't take much to challenge it. You don't have to sweat over a scholarly book as Weber did. You can just blow some pot smoke in the drug czar's face and then go back to your job as a successful capitalist who runs a drug rehab in his spare time to show that not all drug taking is created equal.

IN 1962, HOWARD LOTSOF was nineteen and ready to try nearly anything once, at least when it came to drugs. In fact, inspired by gonzo scientists such as Timothy Leary and Richard Alpert, who, freshly fired from Harvard, were urging people to take mind-expanding drug research out of the industry and university labs and into their own living rooms, Lotsof had set up his own company to manufacture and explore psychedelics. He'd also managed to get himself addicted to heroin. But as just-say-yes as he was, when the chemist who offered him ibogaine told him that the trip could last more than a day, Lotsof decided to pass the dose along to an even more adventurous friend. In the middle of the night a month later, Lotsof got a phone call. It was the friend, raving about the ibogaine.

"It's not a drug," he said, according to Lotsof. "It's food!"

It took Lotsof another six months, but he finally got hold of more ibogaine, enough for himself and twenty of his friends. "No, it wasn't a party," he told me. "Who could imagine a party where everyone was lying around unable to talk? No one would do this for fun."

A nonstop, exhausting series of intense hallucinations, Lotsof's trip covered vast psychospiritual terrain: rebirth ("I dived into a pool, which turned into my mother with her legs open, and I was diving into her vagina"), self-evaluation (a display, like a slide show, of his past life, "all my experiences arranged like files in cabinets"), terror (he was immobilized and unable to stop the images, "the experience so intense and awful I wondered why I had ever done this to myself"), and revelation ("all these thoughts about the symbolism I saw"). All of which was remarkable enough, even to a seasoned psychedelic warrior. But the most significant thing, the thing that changed Lotsof's life, the thing he was least expecting, was that after the hallucinations finally stopped, after he got a few hours of sleep, after he'd gone out onto the Lower East Side streets of Manhattan where he lived, it hit him, stopped him dead in his tracks.

"I suddenly realized that I had absolutely no desire to find or use heroin," he told me. It was the junkie's dream cure: no being strung out, no sweating through the cravings, just the loss of interest, as if all traces of his addiction had simply vanished. And, he found out, four of the six other addicts to whom he'd given ibogaine had the same experience. ("We like being junkies," the other two told Lotsof.)

Lotsof was talking to me by phone from his hospital bed. His leukemia had come back, and he'd just gone through another round of treatment, but he sounded strong and passionate as he told me about everything that was wrong with methadone. He knew this because after he'd been released from jail (he got busted for selling LSD shortly after it was made illegal in 1967), he went to Nepal and promptly got addicted to opium. By then, ibogaine had been put on Schedule I of the Controlled Substances Act, the list of drugs considered too dangerous and of too little medical value to be allowed, and Lotsof instead went into methadone maintenance. A synthetic opioid, methadone staves off withdrawal and craving for as much as a day at a time, instead of the four to eight hours of heroin. This quality is one of the relative virtues of methadone—it frees up addicts to work instead of hunting for a fix—but it also intensifies withdrawal, which is why some call it "orange handcuffs." When Lotsof finally weaned himself, in a monthlong ordeal, he was determined to bring ibogaine onto the market as Endabuse, a remedy that would stop addiction rather than switch it from one drug to a harder-to-kick alternative.

Lotsof has his own drug company, NDA International, which owns patents for using Endabuse to "interrupt addiction," not only to heroin, but to cocaine and amphetamines, alcohol, and even cigarettes. But it's one thing to patent an idea for using a drug and quite another to get approval from the Food and Drug Administration (FDA) to actually market it. The FDA has indicated its willingness to consider an application for ibogaine, but that's an expensive proposition, involving toxicology studies, animal research, and clinical trials. Lotsof has approached virtually every drug company and various government agencies for help but has come up empty. Frank Vocci, who heads the National Institutes on Drug Abuse's Division of Treatment, Research, and Development, thinks he knows why. "If there's something that would make the pharmaceutical industry sprint away from you instead of walking quickly," he told me, "I think it would probably be a hallucinogen."

Indeed, there is no drug on the market that as a normal part of its action causes hallucinations. Add that to the fact that ibogaine is derived from a plant that can't be patented, is intended for a small population that is reviled and chronically underinsured, and is administered only once, and you can see why Big Pharma wasn't eager to pony up the hundreds of millions of dollars it says it spends on developing a new drug simply because an ex-junkie insisted that it was a miracle cure.

Repeated rejection hasn't stopped Lotsof from trying, any more than his leukemia has. He's pestered congresspeople, bombarded regulators, and forced ibogaine samples on scientists, openly desperate to get some help to bring ibogaine into the mainstream of medicine. But he's also worked the fringes. He helped to start the Staten Island Project, a grassroots ibogaine research and publicity effort that grew out of a collaboration with the remnants of the Yippie Party, which in turn wove the failure of ibogaine to attract industry and government money into a conspiracy theory in which the medical-industrial complex's "methadone mafia" was actively conspiring to keep the cure away from addicts and thus maintain the CIA's hard drug business. Lotsof hooked up with the psychedelic psychiatrists who had taken their practices underground when their medicines were made illegal. And in 1989, he set up shop in Amsterdam (ibogaine remains legal there and in many other European countries, as well as in Canada), where he administered ibogaine to addicts in hotel rooms, a service NDA International offered until 1993, when a woman died during the treatment. The cause of the death was never conclusively established, although suspicion centered on the possibility that the patient had used heroin just prior to treatment, but support for the project evaporated—from the Dutch government (and from Erasmus University, where a research study had been under way with the data generated by the forty or so hotel room patients). Lotsof soon relocated his treatments to a clinic in Panama, but by the early 1990s word of ibogaine had escaped his circle. Other people got into the business, like Eric Taub, who said that after an alcoholic friend told him about ibogaine in 1992, he went to Cameroon, scored enough iboga to yield a half-ounce of "pharmaceutical grade" ibogaine, and started dispensing it to addicts and mind explorers alike. Taub, who was once a jeweler, has dosed his clients on yachts in international waters, at Mexican resorts, and recently in a clinic in Barcelona, where he hopes to join forces with a local university at which research on the medicinal properties of ayahuasca, another hallucinogenic plant, is already under way.

Taub has a theory about why ibogaine does what it does to addicts. "It's the granddaddy of all the psychoactive deconstructors of the ego," he told me. "It takes you apart and puts you back together as someone who isn't an addict." Because the drug itself "takes the patient where he needs to go," Taub said, it doesn't really matter that he has no medical or psychological training. (He does require patients to get an EKG and some blood tests, which he said are reviewed by a physician prior to treatment.) In this respect, he's like most of the other providers of ibogaine therapy, which is a thriving cottage industry (he charges upward of $12,000 per treatment). Some of them, like Taub, fashion their own treatment protocol by trial and error, intuition, or whatever they like, and lead people into their interior landscapes like some kind of Indiana Jones of the mind. Others simply provide the drug and a few suggestions for its use. But if you don't want to go totally DIY, if you don't want to guess about things like doses and what to do when someone starts freaking out, you can always do what Marc Emery did. You can download Howard Lotsof's *Manual for Ibogaine Therapy* from the Internet.

THE DROP-OFF AT THE HEAD of the driveway of the Iboga Therapy House is so steep that you have to ignore your certainty that you are about to drive over a cliff. But Marc Emery doesn't interrupt his nearly nonstop chatter, mostly about himself and his life, as he noses the car in. He's done this before. Or maybe he just doesn't take note of precipitous situations.

The house is nestled into a pine and spruce woods, about half-way down the hill from the road to the Sechelt Inlet. The living room sports some comfy furniture and large windows that command a view of the water and the coastline below. Moob looks out one of them while Carriere searches him and his suitcase for crack. She's apologetic but firm, reminding him how important it is that he start clean. She suggests that he change into his pajamas now. Moob glances at the clock. It's just before 5:30 P.M.

Emery is already getting down to business. He has set the kitchen table with a digital scale, a plastic bag full of a yellow-brown powder, and some empty 500 mg gelatin capsules.

"I think around twenty-eight hundred," Emery says.

Carriere looks for a second as if she wants to question him. She says nothing, however, just puts on a pair of surgical gloves and gets to work weighing out the powder and stuffing the gelcaps. Moob is sitting in an easy chair, working a newspaper crossword puzzle. Emery comes out of the kitchen and stands over Moob with five boluses of ibogaine in his hand.

"Twenty-three across is 'Exact,'" Emery says and then hands Moob the pills and a water glass. "Here, man, take this. And I want to tell you something important. When you get up, once it's kicked in, you'll have to move like a robot, like this." He walks a few stiff steps with his arms and legs rigid, his fingers outstretched, head immobile. "Because especially if you move your head, you'll puke for sure."

Moob downs the capsules.

"Twenty-eight hundred is going to be nice and smooth," Emery says. "Smooth in, smooth out."

Moob chases the pills with a long pull of water, makes a face, and complains about the bitter taste. He goes back to his crossword while Emery solves the Jumble aloud. After a half hour or so, Moob's eyes have gone soft and he's cocking his head and squinting at some flowers, as if trying to make sure that they're really there. Emery notices some seals, or maybe they're otters, in the waters below, and Moob makes to get up to get a closer look. He doesn't get more than an inch or two out of his chair before he falls back.

"I think it's time to get into bed," Carriere says.

We adjourn to the darkened bedroom, Moob walking the way Emery demonstrated. Moob gets under the covers, Emery lies across the foot of the king-size bed, and Carriere perches on the side of the bed, near Moob's head. I'm sitting on the floor nearby, and she reminds me that his ibogained senses will soon amplify all stimulation to the edge of unbearable and that my movements when I'm near him should be slow and my speech soft. The four of us make quiet pre-orbital small talk. I mention my dog.

"Does anyone do astral traveling with dogs?" Moob asks. "I'll bet nobody's ever seen a dog do astral traveling." He falls silent.

Carriere signals that it is time to go. She rests her hand lightly on Moob's shoulder. "Remember," she says. "If you see any doors, go through them." She tells him that she will be in and out during the night, but he should feel free to knock on the wall to summon her at any time. "I don't ever want you to think you're asking for too much. If you need a window open one minute and you're cold the next, don't worry about it. I'll close the window. I'll take care of it," she says. Her voice is warm and assuring.

Back in the kitchen, we're joined by Terry, a lanky forty-year-old who doesn't want his last name used in case he decides to go back to work in a normal detox center. He's here to share the overnight patient-minding duties with Carriere. She gives him the particulars, tells him that Moob seems to be well on his way, and they schedule the night's watch duty. Emery then picks up a conversation where we had left it in the car, the one about how he ended up as the Prince of Pot. It's a long tale, running from his days as a teenage comic book entrepreneur in his Ontario hometown, through a couple of wives and innumerable girlfriends, a sojourn in Asia, a couple of stints in jail for drug war–related civil disobedience, and the founding of the mail-order seed business.

Emery did a few more stints in jail and finally reached a truce with the Canadian government, allowing Emery Seeds to prosper and to spawn not only the Therapy House (Moob's treatment, his five-day stay with all meals and ibogaine provided, is free) but also the BC Pot Party, *Cannabis Culture* magazine, an Internet TV station, and, perhaps his crowning achievement: a block of openly pot-friendly businesses (including his party headquarters/bookstore/head shop) in a section of downtown Vancouver where the cops are too busy with the junkies and the crackheads on the streets to bother with a couple of cafés full of stoners chasing their joints with java.

Emery delivers this personal history in an all-news-no-weather deadpan, as if he's told it a million times before—which he might have. Some moments with him do feel entirely staged, like when he spread out a fresh, bright-green pot bud on the income tax return (occupation: "marijuana seed sales") he was showing me. This might be for my benefit, a way to burnish his image as a provocateur and to remind me that he's no champion of sobriety, but he often seems oblivious to me and to others, as if he tells his stories mostly for his own pleasure, as if his autobiography is merely the soundtrack of his life.

Emery, a disciple of Ayn Rand, talks constantly of her everyman-for-himself vision and her exaltation of objectivity as the cardinal virtue, which may be why he is so blasé about the risks he has taken, not to mention the risks Moob is taking right now. I don't know whether this jarring detachment is the cause or the effect of his affinity for Rand or perhaps some secret-identity fantasy grafted onto his childhood fascination with comic book heroes, but his dispassionate self-certainty sounds just like Howard Roark's, a confidence too bloodless to seem arrogant.

It all came together, he told me, when he read *Atlas Shrugged* and "I realized that all those crusading superheroes I'd grown up with weren't just for the comic books. I felt it was my unique destiny to make them real, my duty to go out and change the world."

When it comes to the Iboga Therapy House, even Emery finds his coolness a bit disconcerting. He attributes it to the fact that he's not fighting with cops or prosecutors, so he's not underground. "Ibogaine isn't illegal here, so it can be aboveboard," he says. "It's strange not to have an adversary. For once, I have something everyone wants. Except the drug companies, that is. But they don't even know we're here." Emery talks vaguely about collecting data about his patients, establishing some kind of scientific record about this work, but no one is collecting data tonight. He gets up to take a peek at Moob, who has been completely quiet for the last hour or so. "I don't know. He's not traumatized enough for my liking," he says when he returns. "Twenty-eight hundred might not have been enough."

Emery heads off to the bathroom, and Terry comes into the kitchen. He's been listening in. He tells me that he doesn't think that the off-the-radar approach is enough. "You know, maybe Marc doesn't think we're up against it," he says quietly. "But that's compared to what he's used to. Fact is, we're just a bunch of wingnuts out here. I've been on the inside, I know how far away this whole idea is from what normally gets done. We don't have any degrees. We don't have any research. We don't have anything but balls. Maybe we can just do enough to get someone interested who really knows what they're doing."

AT LEAST ONE PERSON INTERESTED in ibogaine does know what she is doing. Deborah Mash has the degrees, including a PhD in molecular biology, and the research chops—a professor of neurology at the University of Miami, she presides over the Brain Endowment Bank (a collection of brains donated for research by their former owners). Her thick résumé documents a brilliant career that includes the discovery of cocaethylene, a poison that the body manufactures out of cocaine and alcohol, which accounts for the devastating effects of that combination of drugs. And she agrees with Terry about Emery and Taub and all the other underground ibogaine providers.

"They're wackadoodles," she said, "and they're going to ruin this for everyone."

Mash, who is forty-seven, met Howard Lotsof when he approached her for information about the relationship between cocaethylene and the action of ibogaine. "He knew nothing about neuroscience," Mash told me. "But I thought, if this is real, why is no one looking at it? I thought we had a responsibility as medical researchers to test it. So I went to Amsterdam with Howard. He wasn't collecting any data. I said to the patients, 'Okay, boys, piss in the cup,' FedEx-ed the samples back, and got some very interesting preliminary data, enough to get me interested. Of course," she added, "I had no idea what I was getting myself into."

What she got herself into was Lotsof's Endabuse business. And while that alliance has perhaps nudged ibogaine closer to scientific respectability, yielding some journal articles on its pharmacology and efficacy, it has also led to protracted and bruising battles with government regulators, fellow scientists, and ultimately with Lotosf himself, who sued her for patent infringement. (After years of litigation, the case was settled; neither side can discuss the terms.)

Even if she wasn't expecting it, though, Mash seems to enjoy a good fight. Like Marc Emery, she's a crusader, but if a conversation with Emery is a smooth cruise on a four-lane in the prairie, talking with Deborah Mash is a white-knuckle careen down a switchback mountain road. A trim and intense woman who seems to be moving even when she's standing still, she is at one moment reeling off facts about the *pharmacodynamic profile of ibogaine in the synapse* and *kappa-opioid receptors* and *medical inclusion criteria* and tearing up the next, her voice thick with emotion as she says, "Gary, I wish you could have been in the meeting today—I wish to God you could have been in that meeting. When the clients left, the staff sat there and basically prayed. We were so blown away by what we heard." One moment she's defending the government agencies like the National Institute on Drug Abuse (NIDA) that have made her life so difficult: "I know that NIDA has a very broad agenda, from HIV to women's health to children to fundamental neuroscience. They've got a lot of areas that they need to cover." And the next she's railing against the bureaucracy and lamenting her martyred career: "My colleagues have hung me out to dry."

And, she tells me, one moment she's weeping over getting rejected by NIDA for the money, the measly hundred thousand dollars or so, that she needs to conduct a study, already green-lighted by the Food and Drug Administration, to test ibogaine in human subjects (the first step toward approval for the drug), and the next she's resolving to do exactly the opposite of what a main-stream scientist concerned about her reputation does: move off-shore. On the heels of the NIDA rejection, she rounded up some investors, worked some international contacts, and established Healing Visions Institute on the Caribbean island of St. Kitts, a for-profit ibogaine clinic where she treats people and uses them as guinea pigs in her own private drug trial.

St. Kitts has a sputtering sugar industry and, thanks to its rocky coastline, limited tourist prospects. Mash and I are the only dinner-hour customers in one of its few decent restaurants right now, going over the ground rules for tomorrow's visit to the clinic: no real names, not even first names; no talking to the patients with-out a staff member present; no entry to the medical building when patients are present; no staff interviews without her permission.

Even if they weren't standard-issue restrictions on report-ing about medicine, I would know them by heart now; they've been part of virtually every conversation Mash and I have had since we started talking about my visit four months ago. She has been a tenacious and unpredictable negotiator. She's summoned me to a Sunday afternoon conference call with her lawyers, can-celed one trip as I left for the airport (a celebrity had signed up for treatment at the last minute, she explained much later), and revoked permission for another upon discovering that I had called the St. Kitts minister of health to get his view of Healing Visions, a move that she warned could set off an "international incident" and ruin her. (The minister who alerted Mash to my call but never returned it turned out to be her local medical director.) But I've wheedled my way back into her good graces and am now, as she keeps reminding me, the first journalist to be granted access to Healing Visions.

Which turns out to be an unspectacular and underused resort, a cluster of pastel-colored stucco-and-frame cottages perched on a dry scorched hillside above the Atlantic Ocean and surrounded by a barbed wire–topped cyclone fence. Mash says the fence was already there when she started renting the place, which she now does six times a year for ten-day sessions, each attended by between eight and fifteen patients. Healing Visions is every bit Mash's show. When she walks through the pavilion where meals are served, the patients hush their conversations and move aside like parting waters. The staff of about a dozen doctors, nurses, and counselors brings matters of every significance to Mash. In a single half hour she counsels a patient, schedules a staff meeting, brainstorms with a physician about a medical problem, takes a phone call from a prospective patient, talks with a doctor on the mainland about a former patient, and deals with a balky washing machine. She says that she rarely sleeps more than a few hours a night during the sessions.

"She's incredibly dedicated," one Healing Visions alumna told me. "She made a million phone calls to convince my family that it was an okay thing to do. You know, you can't just say, 'Can I have twelve thousand dollars to go to St. Kitts and take a hallucinogen that's not FDA approved, please?'"

Not everyone appreciates Mash's intense, sometimes abrasive, style. "She's like a used car salesman," an ex-patient told me. But he quickly added that this didn't stop him from going to Healing Visions—or from kicking the codeine and alcohol habit he'd had for more than a decade. If she's a control freak, it's because the situation demands it, not only because the lives and well-being of her patients (not to mention her own scientific career) are at stake but also because there is a precedent for using hallucinogens as therapy for addiction, and it's not a pretty history. In the 1950s, after the Central Intelligence Agency abandoned its attempt to develop LSD as a weapon—a mist, perhaps, that would disable the Russians—the drug was used by doctors in Canada and the United States to treat alcoholics. The results were remarkable: single treatments led to long periods of relapse-free living at much higher rates than other therapies. But then the drug escaped the labs—a catastrophe, according to Mash.

"Timothy Leary did so much damage," she says over dinner. "I'm not going to re-create that." That means, of course, no drugs for her staff or for her. Not a problem, she says, because "I've never met a drug I like. My spiritual core is solid, my relationship to my frontal lobe is great."

But the medical jargon that Mash constantly uses with staff and patients alike—the insistence on being called "Dr. Mash," even though she is not a physician; the blood pressure cuffs and phlebotomy kits and urine containers in the treatment cottage; the four-inch-thick, FDA- and NIDA-friendly dossiers she amasses on each patient; the constant insistence that ibogaine only gives people a head start on their journey into recovery from their chronic illness—all these reminders that this is a *medical* clinic, not some spiritual retreat for druggies, can't overcome a glaring fact: the stories the patients tell are not about disease and its cure but about spiritual decay and transformation.

Roberto, for instance, may have had an IV inserted into his arm for monitoring blood levels and been attended by doctors and nurses, but the day after he got his dose, this tanned and shirtless and heavily tattooed young man isn't talking about anything remotely medical. A twenty-four-year-old veteran of countless failed detox attempts, he's positively radiant as he tells me that he's a seasoned consumer of psychedelics, that when it comes to hallucinations he thought he'd seen it all, and that ibogaine proved him wrong.

"It just kind of took over my body," he tells me. "Grabbed me by the throat really." His big smile fades as he describes the visitations from his girlfriend and the grandmother who raised him, both of whom are dead and both of whom he felt he had let down.

These visions made clear to him that "my life has been about me, myself, and what I could do to get something from you."

It wasn't all rebuke, however. "I felt guilty for my girlfriend's death—she OD'd—and it was like she was telling me, 'Don't feel guilty. I'm all right, and it's okay to move on with your life.'" And then he was transported to a beach, where his grandmother was holding him and "all of a sudden, out of the water, this big angel came out. It was like Mother Earth, like my creator, something higher than me letting me know everything was going to be all right." Roberto stops, overwhelmed. The counselor sitting with us reaches across the table, puts her hand on his arm. "I can't say more," he says. But he keeps trying. He wants me to understand how different this feels from every other time he's sworn off heroin. He explains how he'd come to the island half-willingly and fully skeptical, more so that he could say he was trying than to actually get clean; how he was a callow young man who had never been serious about quitting before, and that the vision had entirely changed his perspective. "I don't know how to explain it. I got in touch with the kid who got lost between the drug world and the insanity of my life. And now I'm more at peace." Roberto is crying now.

The counselor, still touching Roberto, turns to me and tells me about his aftercare program, how they will coordinate further treatment with his doctors and his family (he lives in a city far from Miami) and will urge Roberto to attend group meetings. Roberto sounds a little defensive now. "I'm gonna do all that. I know I still have to fight my addiction. But now I feel like I got a foundation," he says, as much to her as to me.

At moments like this, you really feel the clanging juxtapositions of a place like Healing Visions and the difficulty that Mash faces as she tries to straddle two worlds. It happens when you talk to her as well. For all her insistence on scientific data, all of her repeated avowals that ibogaine isn't a magic bullet cure, that addiction requires a lifetime of recovery (and everyone goes home from Healing Visions with an aftercare program that includes attendance at twelve-step groups), she's never far away from another anecdote about a dramatic transformation. The man who "died over and over and over again, the ibogaine saying, 'Here is what death is, see it? See it?'" The woman whom the ibogaine showed two images: "one was a coffin with her two-year-old in it and the other was a twelve-year-old beautiful young man. That's her choice, right in front of her face."

Mash tells these stories with a ferocious conviction that drug addiction is a debased state, a spiritual corruption, and when she reverts to her science talk, you wonder whether she is protesting too much, working too hard to be the anti-Leary, going out of her way to make sure no one gets the wrong impression: that she thinks that addiction is moral, rather than medical, or that anything as unscientific as a drug trip could change people once and for all.

THEY DON'T EVEN GIVE LIP SERVICE to the medical model at the Iboga Therapy House. Moob got the tests that were suggested in Lotsof's manual and gave Emery the results, but I'm not sure he even understands them. And Moob's medical and psychological history is mostly unknown, there hasn't been any talk of aftercare, and there probably won't be. Right now, the morning after a very quiet night (I heard him speak once, to complain that he shouldn't have had that second glass of water as he robot-walked to the bathroom), he's not ready to do that.

"I just can't," he says. "Too much came up too fast."

But he's clear on one point: the last thing he wants to do right now is take a puff on a crack pipe. He'll have a few days to hang out here, talk about his experience with Terry and Carriere if he wants, and maybe fish off the dock before going back to the city to test his resolve.

Emery has no interest in bringing ibogaine into medical respectability. He would prefer simply to operate his house, add the plant to his catalog, and sell addicts the seeds of their own cure (along with instructions, which he already supplies for his pot seed customers), so that he doesn't have to wrestle with the implications of ibogaine for orthodoxies about addiction and its treatment. Indeed, to him the fact that ibogaine is a heresy is part of its attraction.

But Deborah Mash doesn't have the luxury of indifference to the medical mainstream, and she has an idea that might get ibogaine, or some variant of it, past the gatekeepers. She has isolated a metabolite of ibogaine—noribogaine—that she says is responsible for the long-lasting anticraving effects of the drug. Unlike the plant itself, noribogaine can be patented and, Mash thinks, turned into a treatment much less paradigm-busting than what's going on at Healing Visions: a patch that would release a steady drip of the chemical into the addict's blood, staving off desire for drugs in the same way that a nicotine patch does. Noribogaine also has one other quality that makes it attractive: it appears to be nonpsychoactive, preventing withdrawal symptoms and inoculating against craving in the deep biochemical background, without the patient's awareness and, most important, without causing hallucinations.

The office is unusually quiet right now, with the counselors either attending to the people taking their ibogaine trips today or presiding over the group meeting where other patients are telling their stories of yesterday's journeys. The phone is quiet, and Mash, her data files stacked on the counter behind her, is explaining noribogaine to me and is suddenly, disconcertingly, equivocal about the relationship between the visions and the healing.

"I still don't know the value of this content," she says. "Certainly, it's striking, but maybe it's not as important as we think. We can still do the ibogaine detox, allow the visionary piece to be there, but maybe it wouldn't be necessary. Maybe noribogaine would be enough to give them that window of protection as they enter treatment or go back into the workplace. I think that model is a winner." Maybe, in other words, the visions are a mere side effect of a neurochemical storm that leaves an unaddicted brain in its wake. "Like a little chemical ECT [electroconvulsive therapy] resetting neurotransmitter systems," Mash ventures.

Stanley Glick, the director of the Center for Neuropharmacology and Neuroscience at Albany Medical College, agrees. "It's almost like a reboot," he told me from his office. Glick has been investigating ibogaine for nearly half of his thirty-five-year career. He was drawn in, like Mash, by Howard Lotsof, who he thought was "a complete lunatic." But Lotsof was persistent, and partly out of curiosity and partly to get rid of Lotsof, Glick decided to give some ibogaine to the morphine-addicted animals in his lab.

"When you hear the same amazingly similar story enough, you think there must be some truth to it," he said. "I figured I'd take a look at it and a couple weeks later I'd be done with it."

But the drug decreased the rats' self-administration of morphine. So Glick started to synthesize compounds that had a similar structure, and when some of them turned out to have the same effect on his animals, "I was hooked."

Glick eventually discovered and patented a drug he calls 18-MC. It reduces morphine intake in his animals, and when he takes away the morphine completely, the animals don't seem to have withdrawal symptoms (which can be observed in rats through behaviors like shaking and torpor). Glick's drug also appears not to depress the cardiorespiratory system, and, perhaps most important, it is almost certainly not psychoactive. Glick even has a theory about why this happens: 18-MC is active mostly at nicotinic receptor sites, an integral part of one of the brain's main reward pathways, but not at the serotonin sites, which are implicated in hallucinations. By binding to the nicotinic receptors, he said, 18-MC suppresses the neurons' clamoring for the excitation that morphine—and, Glick said, just about any addictive drug, including nicotine and alcohol—would provide.

Despite these promising results, Glick has yet to raise the $600,000 or $700,000 he estimates it will take to do the preliminary toxicology studies that the FDA requires before it will approve the drug for testing in human subjects. It seems that expanding ibogaine's target population, elucidating its neurochemistry, and distilling its antiaddictive from its hallucinogenic properties still aren't enough to overcome the drug industry's resistance to 18-MC. But here Glick thinks he's up against a more traditional roadblock than the one he would face with whole ibogaine, which he thinks doesn't stand a "ghost of a chance of going anywhere in this country."

"There's virtually no other drug that works by this mechanism," he told me. "This is a very conservative industry, perfectly content to invent new ways of doing the same thing, but they're very reluctant to do anything different."

Tell an ibogonaut that the visions are incidental, that the journey was perhaps no more than the brain occupying itself during its wipe-and-format, and you're likely to be met with an incredulous stare. "Bullshit. I know what happened to me," Sheldon said. He gestured to the Vancouver street. "If they'd just put me to sleep or something and hit me up with it, I'd still be out there using." When Terry heard that this is what the people who know what they are doing are up to, he looked stunned. "Really?" he asked. "Isn't that a little like burning the village to save it?"

But back in the United States anyway, there are only two kinds of drugs: the kind your government has decided are good for you and the kind that are so bad you'd be better off in jail than using them. Ibogaine may have to be domesticated, its visions turned into side effects and excised, for it to move across that border.

A drug that can reboot the brain without the messy complications of altered consciousness or that can drip resolve instead of heroin into an addict's veins is an obvious winner, not only because it fits the drug industry's paradigm so much better than a hallucinogenic plant from Africa or even because it leaves intact all that we have come to believe about addiction as a chronic illness. Distilling the healing from the visions also adapts ibogaine to the current fashion in understanding ourselves: that the despair of addiction and the transcendence of getting clean, indeed all of our troubles and triumphs, are just so much neurochemical noise. This is, after all, what it means to call addiction or any other complex set of behaviors a disease: that it will be located in our flesh, ultimately in our molecules, and that consciousness is the biggest side effect of all.

2

DEPRESSION:
IN THE MAGIC FACTORY

D<small>R. GEORGE PAPAKOSTAS HAS SOME</small> bad news for me. For the last half hour, he's been guiding me through a catalog of my discontent: the stalled writing projects and the weedy garden, the dwindling bank accounts and the difficulties of parenthood, the wife I see mostly in the moments before sleep or on our separate ways out the door, the bad dreams and the stink of mortality that flee the room as I wake up from them, the typical plaint and worry and disappointment of a middle-aged, middle-class American life that you wouldn't bore your friends with, that you wouldn't bore yourself with if you could avoid it, and if this sweet man with his solicitous tone hadn't asked. He's been circling numbers and ticking boxes, occasionally writing a word or two in the fat three-ring binder on his desk, and now he's stopped the interview to flip the pages and add up some numbers. His brown eyes go soft behind his glasses. He looks apologetic, nearly embarrassed.

"I'm sorry, Greg," he says. "I don't think you're going to qualify for the study. You just don't meet the criteria for Minor Depression."

Even if my confessor had gotten my name right, I'd be a little humiliated. I'd come into his office at the Depression Clinical and Research Program of Massachusetts General Hospital, a consortium of Harvard and two teaching hospitals, insisting that I would qualify. I told him that I figured anyone paying sufficient attention was bound to show the two symptoms of Major Depressive Disorder (out of the nine listed in *DSM-IV,* the latest edition of the American Psychiatric Association's *Diagnostic and Statistical Manual of Mental Disorders*—sadness, diminished pleasure, weight loss or gain, trouble sleeping, fatigue or malaise, guilt, diminished concentration, and recurrent thoughts of death), in sufficient quantities to cause distress, that are required for the Minor Depression diagnosis. To explain my certainty and my interest in his study, I had recited some facts: that these days my native pessimism was feasting on a surfeit of bad news: my country taken over by thugs, the calamity of capitalism more apparent every day, environmental cataclysm edging from the wings to center stage, the brute facts of life brought home by the illnesses and the deaths of people I love and by my own creeping decrepitude. That I'd more or less resigned myself to my dourness, that it struck me as reasonable, realistic even, and no more or less mutable thany short stature, my constitutional laziness, my thinning hair, my modest musical talents, the quirks of personality that drive away some people and attract others. That I consider melancholy not so much an illness but a boundary condition that, even if I regret it at times, still leaves me with a decent life, played perhaps in a minor key, but still better than most people have or anyone deserves. That I'm not *undistressed* by these conditions, that I can think of many other arrangements, designs that strike me as more intelligent—a nature, for instance, as generous with pleasure as with pain—or at least more humane. That as a therapist, I lean toward talk therapies for psychic distress, but I am not at all opposed to better living through chemistry, so if the drugs offered by his clinical trial (Celexa, Forest Laboratories' blockbuster antidepressant, and Saint-John's-wort, an herb with a reputation as a tonic for melancholy) did what they promised, I might like that, and if I did not, at least I'd know what I was turning down. And finally, that I was going to write about whatever happened, which meant that either way I wouldn't come away empty-handed.

Unless I didn't *meet the criteria*.

But before I can get too upset, Papakostas has more news. "What you have is Major Depression," he says. He looks over the notebook again. "It's mild, but it's not minor. Nope. Definitely Major Depressive Disorder with Atypical Features, Chronic."

Which means, he seems pleased to tell me, that I meet the criteria for four or five other studies that Mass General is running. I can take Celexa or Mirapex or Lexapro or something called S-adenosyl-L-methionine. I can climb into an MRI, get my brain hooked up to an EEG, and take home a device to monitor my pulse and breathing. I can get paid as much as $360 for my trouble. I can go back to the waiting area, read over the consent forms, which spell out in great detail—down to the final disposition of the two tablespoons of blood that they will take—all that will happen to me, what is expected of me, what my rights are, how I can bail out if I want to, and then make my decision.

I'm a quick shopper, and when Papakostas returns, I have already signed the papers for Research Study 1-RO1-MH74085–01A1, agreeing to return next week and then every other week for the next two months so that they can evaluate the alleged antidepressant properties of omega-3 fatty acids—fish oil. (The rationale for studying fish oil is that according to the World Health Organization, the countries with the highest consumption of fish have the lowest rates of depression. And it happens that omega-3s make cell membranes, such as the receptors in your brain that absorb serotonin and other neurotransmitters, more supple. To a psychiatrist already convinced that depression is the result of deficiencies in serotonin transmission, the significance of this correlation outweighs any of the other possible explanations for why someone in fish-deprived New Zealand or France might be more depression prone than someone in Korea or Japan.)

Which is why right now Julie and Caitlin, freshly minted college graduates, are hovering over me in a tiny exam room, just a metal table and a scale and a phlebotomist's chair, tweezing tentatively through the thatch on my chest, and worrying out loud that they are hurting me. They finally clear the spots for the EKG electrodes and run the scan of my heart. They take my pulse and blood pressure, weigh and measure me, and draw—with barely concealed trepidation—my blood into a syringe. Fair-skinned Caitlin is blushing a little as she hands me the brown paper bag with a cup for my urine specimen. I can see how cowed these young women are by this forced intimacy, and I try to tell them they needn't be so shy. But they know I have just been declared *mentally ill,* and I wonder whether reassurance from the likes of me only makes things worse.

I haven't come here to minister to them, however, or, for that matter, to maintain my dignity. In this nondescript office building hard by the towers and pavilions of Massachusetts General Hospital in Boston, these dedicated people do the research that determines whether drugs work, which is to say whether they will come to market as government-sanctioned cures. In the process, they turn complaint into symptom, symptom into illness, and illness into diagnosis, the secret knowledge of what really ails us, what we must do to cure it, and who we will be when we are better. This is the heart of the magic factory, the place where medicine is infused with the miracles of science, and I've come to see how it's done.

I NEVER USED THE TERM "magic factory"—you wouldn't want to seem *paranoid* in a place like this—but I told Papakostas about my suspicions of the drug industry and even referred him to what I'd already written on the subject. If he caught a whiff of bad faith here, if he thought me like a bluestocking on an evidence-gathering excursion to the porn shop, or if he worried that I would lie to him just to get a story (he knew I was a therapist, that I was intimately acquainted with that punch list of symptoms), he was too good-natured to say so. (Or too hard up for subjects. The investigators expect that it will take five years to enroll the three hundred subjects needed to complete the study.) But then again, he's a doctor and has to believe that if depression is the medical illness that the antidepressant industry is built on—if it is, as the drug company ads say and as doctors tell their depressed patients, a chemical problem with a chemical solution— then my intentions shouldn't matter. Diseases don't care whether you believe in them. What matters is the evidence, how much insulin is in the blood or sugar in the urine and all the other ways nature has of telling you that something's wrong.

But there's no lab to send my bodily fluids to in order to assay my levels of depression. Instead, there are tests like the 17-item Hamilton Depression Rating Scale. The HAM-D was invented in the late 1950s by a British doctor, Max Hamilton. He was trying to find a way to measure the effects of antidepressants that the drug companies were just bringing to market. To figure out what to test for, he observed his depressed patients and distilled their illnesses into seventeen items, like *insomnia* and *feelings of guilt*. Patients could get as many as four points per item, and eighteen of the fifty-two possible points is considered the threshold for depression. Ten of the seventeen items were about neurovegetative signs such as sleep and appetite, the kind most likely to respond to antidepressants—something Hamilton knew because he'd worked with the drugs. Not surprisingly, this drug-friendly test quickly became a favorite of drug companies. In fact, it remains the gatekeeper to the antidepressant industry, used universally by the Food and Drug Administration to evaluate candidate drugs.

But because Max Hamilton had already decided that the people from whom he derived the items were sick, the HAM-D cannot be used to *diagnose* depression. A person with a high score doesn't have depression any more than a person with a sore throat and a fever necessarily has strep. So psychiatrists have developed a diagnostic test, the Structured Clinical Interview for DSM-IV (SCID), which is tied to the *DSM-IV*'s catalog of the afflictions that cause people sufficient "psychic distress" to be considered official mental illnesses. (The second edition of the *DSM*, published in 1968, lists a mere 168 diagnoses.) The *DSM-IV*'s fifty-one possible mood disorders take up 84 of its 943 pages, which list criteria and specifiers that a clinician assembles into diagnoses like Major Depressive Disorder with Melancholic Features, Chronic with Seasonal Pattern. If *DSM-IV* is the *Audubon Guide* to psychic suffering, the SCID is the psychiatrist's Baedeker, guiding his venture into an unknown country and making sure he'll find all the birds he's after.

There's no magic to the test. To determine whether you meet the *DSM-IV* criterion of "depressed mood most of the day, every day," it asks, "Has there been in the last month a period of time when you were feeling down most of the day nearly every day?" To find out whether you have a "diminished ability to think or concentrate," it asks, "Did you have trouble thinking or concentrating?" And so on with all the lists of symptoms, until, based on your answers, you get shunted, like coins in a sorter, from one chute to another until you drop into the drawer with all the other pennies.

I never saw the scoring from my SCID, so I'm not sure exactly how I ended up with my diagnosis. (It was a good thing, however, that it would not be entered into my medical dossier, where it might wreak havoc on future attempts to get life or health insurance or run for president.) I do know that I told Papakostas the truth, at least to the extent that I could figure out how to answer his questions about my psychic life. And I also know that in the course of a quarter-century of practicing therapy, I have met people who are hammered flat, unable to find solace in any quarter, their self-loathing turning sweetness into ashes in their mouths, people who are nearly insensate to anything other than their abject misery, who can think of little other than dying, who, in short, meet the criteria in my own private *DSM* for Major Depression: a handful of them, maybe ten or twenty out of the seven hundred or so patients I've seen. It was hard to believe that whatever my score on the SCID, Papakostas really thought I was majorly depressed. I wasn't tearful with him, and while I whined about the things that the SCID invited me to whine about, I was alert and smiling, joking, digressing, and more effusive—perhaps out of nervousness—than I normally am.

I didn't say this to Papakostas, didn't protest that I'd met and known Major Depression for many years, and my aches and complaints were no Major Depression. Just as well: Item 17 (Insight) on the HAM-D awards two points for "denying being ill at all."

IT'S NOT JUST MY OWN private *DSM* that wants people to be virtually disabled before they qualify for the diagnosis. The official book is very clear: if you really meet the criteria, you very likely won't be getting out of bed to get interviewed by a psychiatrist, let alone bantering and exchanging professional chitchat with him. Hamilton's subjects were already in the hospital, and early research was done primarily with inpatients, who are generally much more debilitated than the walking wounded. But the industry's appetite for "depressed" subjects is as unlimited as the market's appetite for novelty, and over the years, clinical trials have come more and more to depend on outpatients.

When you talk to psychiatrists about this, their forthrightness is disconcerting. Donald Klein, a pioneer in psychiatric drug research and a renowned Columbia University physician, once told me, "The problem with antidepressant studies is that anything that can be confused with ordinary unhappiness gets in." Lawrence Price, who directs research at Brown University, pointed out, "If you go out and advertise in the newspaper for depressed people, you are going to get less-ill people than if you are taking people who are brought in via the emergency room. And if the investigator has directed his or her research assistant to rate liberally on the Hamilton, then you are going to have more people meeting the entry criteria." At $10,000 per subject, Price added, investigators are highly motivated to exaggerate Hamilton scores. A colleague of Price's at Brown has estimated that the scores are inflated by up to five points for clinical trials.

This is an open secret among researchers, but they're not telling the rest of us, and they probably won't until someone figures out what to do about it, at which point the fix will be announced as a major breakthrough. But even if you have a killer app for your drug, the kind where there's no question who fits the diagnosis, you're not necessarily in the clear, especially if the drug falls on the wrong side of a deeply held belief, even a fictitious one. That's what would-be entrepreneurs of ibogaine have found out: that because addiction is an allergy to *any* drug, the idea of treating it with another drug is a nonstarter.

There are ways around this problem, too. Naltrexone, for instance, a drug that blocks the action of opiates, or methadone, a drug that acts primarily to quell the craving for heroin. These drugs most likely earn their exception from the no-drugs-at-all rule because they fit so nicely into the pharmaceutical model: they are taken every day, and perhaps most important, they don't get you high. Which is not something you can say for ibogaine. Nor, for that matter, can you say it for marijuana, but that hasn't stopped an English company, GW Pharmaceuticals, from developing a drug called Sativex for the treatment of multiple sclerosis. Sativex is nothing other than pure cannabis, a blend of the same high-potency strains that Marc Emery smokes rendered as a liquid that is sprayed under the tongue. Patients will tell you that the drug gets you good and stoned, but GW describes Sativex as a "novel prescription pharmaceutical product derived from components of the cannabis plant." The company claims to have distilled the medicinal from the recreational properties of cannabis and to have reduced intoxication to a "side effect." They have hired Andrea Barthwell, who once spearheaded the American drug czar's campaign against medical marijuana, to spread this news, which even GW's own researchers will tell you, if you ask, is "a load of bollocks." And the FDA has put Sativex on the fast track to approval in the United States.

GW understands that medical fictions, noble and otherwise, can be used to your advantage if you're clever enough. Sativex stays on the right side of the mythical divide between medicinal and recreational drugs. Antidepressant manufacturers don't have to steer quite so skillful a course as GW because the fact that their drugs were discovered as Viagra was—people taking them for one purpose found themselves enhanced in an unexpected way—has long been obscured. But they've got something else on their side besides historical amnesia. They've got a disease, depression, that is tailor-made for their product.

JULIE GREETS ME when I arrive the next week. I'm eavesdropping on the receptionist, who is reassuring someone on the phone that many of the doctors at the Depression Clinical and Research Program teach at Harvard. I get my medicine today, assuming that my EKG checked out and that my blood and urine were clean of illicit drugs and any indication of disease. Julie hands me a clipboard with three questionnaires and a pen. The Quick Inventory of Depressive Symptomotology, the QIDS-SR, is sixteen multiple-choice questions. Here's Item 11:

View of Myself

0. I see myself as equally worthwhile and deserving as other people.
1. I am more self-blaming than usual.
2. I largely believe that I cause problems for others.
3. I think almost constantly about major and minor defects in myself.

The Q-LES-Q, the Life Enjoyment and Satisfaction Questionnaire, wants me to circle the number from 1 (Very Poor) to 5 (Very Good) that describes how satisfied I've been during the last week with sixteen aspects of my life, from my economic status to my sex drive, interest, and/or performance. And on the Ryff Well Being Scale, I can express, by filling in the little bubbles, as on an SAT, one of six degrees of agreement with fifty-four statements about my attitude toward life, such as, "For me, life has been a continuous process of learning, changing, and growth" or "My daily activities often seem trivial and unimportant."

The psychological tests in the women's magazines dotting the waiting room tables aren't much different from these, save for one thing: social scientists have stamped their approval on the official questionnaires after subjecting them to various statistical challenges and worrying over such things as the fact that people will answer self-report inventories according to how they want to look to the tester. But aside from a passing frisson over telling Julie, however elliptically, about my *very good* sexual performance, I am not thinking about impressing her. I'm thinking about how little I seem to know about myself. I didn't know, for instance, that wondering whether "life is empty" or "if it's worth living," as I do at least once a week—often, I notice these days, when a patient of mine tells me about the illness or death of someone they love; I can catalog all the ways in which lives can crash and burn and sear other lives in the process—is, as the QIDS insists, a "Thought of Suicide or Death." I think I march to my own drummer just as much as the next guy, but when the Well Being Scale asks me to rate how difficult it is "for me to voice my own opinions on controversial matters," I think of how I often find myself disagreeing with myself about what my opinion is, how the closer I get to fifty the less sure I feel of anything, even the answer to this question, and I can't find a place to bubble in that uncertainty. I wonder what it means that I hesitate so long over these questions, whether I should circle the QIDS item that says, "My thinking is slowed down."

Slowed down or not, I'm not finished with the Well Being Scale when Papakostas comes to fetch me. He's still calling me Greg. I tell him I'm confused about a consent form Julie just handed me, explaining that the one Papakostas and I signed last week was "outdated." But, I tell him, this new form seems to be for a different study, one that seems to require me to take two different pills at the same time. He looks perplexed, excuses himself, and returns with Julie. Together, they explain that the study I signed up for last week was full, so they reassigned me. He looks mortified. Julie, who told me she was fresh out of Amherst, looks worried. They're explaining, apologizing, and reassuring, as if they were waiters in a restaurant who have just delivered the wrong meal to a valued customer.

But I'm not complaining. I'm not the least bit discomfited, except perhaps at the fact that a doctor surrounded by the accoutrements of his eminence, all the professional bric-a-brac on his walls, the eager young assistant, the prestigious hospital outside his window, a position from which he wouldn't have to issue more than a muffled apology if he cut off the wrong leg, is now fawning over me about a fuckup in the paperwork. On the other hand, we all know what has happened here. They've broken the code, the Nuremberg Code, the one that says that you can't do human experimentation unless the human in question knows exactly what he's getting himself into, of which it is their responsibility to fully inform me. Not only that—which is bad enough, since the U.S. government is paying $2.5 million for this research, funding that is contingent on paying scrupulous attention to such matters—but for a moment they've laid bare the thing that all this scrupulous attention to my autonomy is supposed to obscure (because, of course, it can't be eliminated, it is the whole point): that they are using me, that my Well-Being, my Life Satisfaction, my blood, and my piss will all get rendered into raw data for these doctors and for other doctors running other trials for other drugs, seeking the thumbs-up that injects them into the marketplace or the thumbs-down that sends the scientists back to the drawing board, with investors following behind like so many lemmings. They've moved me around like a pork belly, and for a split second, the bald fact of the commerce we are conducting, and of our respective roles in it, is right in front of our faces.

So I reassure them that I'm satisfied with their disclosures, that I just wanted to make sure we were all on the same page. Julie leaves the room with a last apology, and Papakostas hands me back my copy of the form, countersigned by him. He opens the binder again and asks me how my week was. Papakostas has a way of cocking his head and holding me in his gaze and of making the HAM-D into a reasonable facsimile of an actual conversation. So when he asks me for an example of what I feel self-critical about (Item 2), I open the spigot a little, tell him that I worry that my insistence on working at my practice part-time, my giving up a plum teaching job, my indulgence in writing and other even less savory vices, my seemingly endless desire for free time—that these reflect a constitutional laziness, a hedonism, and an irresponsibility that have led me to squander my gifts. Papakostas waits a beat, then nods and says, "In the past week, Greg, have you had any thoughts that life is not worth living?" It's time for Item 3.

Papakostas is so unfailingly kind—and I want him to *care,* I want him to tell me that I am not really feckless—that I can't be mad at him for this, let alone correct him about my name. He's not doing it because he's a bad man or a disingenuous one or a shill for the drug companies but, on the contrary, because he does care. He thinks I am suffering; he is a doctor, and this is what he knows how to do: to find the targets and send in the bullets, then ask the questions and circle the numbers, and decide whether the drugs really are doing their magic. We're not here to talk about *me,* at least not about the dodgy homunculus we call a self. We're trying instead to figure out what is going on in my head, literally, in the gray, primordial ooze where thought and feeling, according to the latest psychiatric fashion, arise.

BACK ON THE STREET, blinking in the noon sun, I peek into my brown paper bag. The study medicine comes in a pair of plastic bottles stuffed with two weeks' worth of glistening amber gel caps. They look just like regular prescription drugs, except for the sticker that says, "Drug limited by federal law to investigational use." That seems a little dramatic for something I can get at any health food store or by eating however much salmon it would take to provide two grams of omega-3s per day. But under the agreement we've made—that they are doctors, that I am sick, and that I must turn myself over to them so they can cure me—the medicine must be treated with the reverence due a communion wafer.

Not that anyone at Mass General would say so. In fact, they've designed this study to minimize the possibility that something as unscientific as faith or credulity or the mystifications of power could be at work here. The trial is a so-called three-armed study. I have been randomly assigned to one of three groups. One group gets placebos in both bottles. One group gets eicosapentaenoic acid and a placebo, and the third group gets docosahexaenoic acid and a placebo. Only the anonymous pharmacist laboring in the bowels of Mass General, armed with a random number generator and sworn to secrecy, knows which group I'm in. The study will then be able to show which of the two omega-3s has more effect, and whether either one is more powerful than a placebo.

This method is known as the double-blind, placebo-controlled design and provides a way to deal with something that the drug industry would rather forget: that in any given clinical trial, especially for psychiatric drugs, people are very likely to respond to the fact that they are being given a pill—any pill, even one containing nothing but sugar. Which is why the FDA requires all candidate drugs to be tested against placebos, to try to separate the medicine from the magic, to see what the drug does when no one is looking. But like a pain-in-the-ass brother-in-law, the placebo effect keeps showing up at the drug companies' parties, curing people at a rate that is alarming to both regulators and industry executives. In fact, in more than half of the clinical trials that were used to approve the six leading antidepressants, placebos outperformed the drugs. In addition, when it came time to decide on Celexa, an FDA bureaucrat wondered on paper whether the results were too weak to warrant approval, only to be reminded that all the other drugs had been approved on equally weak evidence.

Despite the fact that the placebo effect is the indirect subject of virtually every clinical trial, no one really understands how it works. Science, which was designed to break things down to their particulars, can't detect something so diffused through-out the encounter between physician and patient, so ineffable. Until there's money to be made in sugar pills—at which point the drug companies are sure to investigate it thoroughly—about the best we can say is that the placebo effect has something to do with the convergence between the doctor's authority and the patient's desire to be well. But this relative ignorance doesn't stop doctors, wittingly or not, from using their power as a healing device. For instance, they can reshape you in a way that makes you a good fit for the drugs. That's what the diagnostic tests, with their peculiar method of inventorying personhood, do: they alert you to what it is in yourself that is diseased—casting your introspection as *excessive self-criticism,* your suspicions of your own base motives as *low self-esteem,* your wish to nap in the afternoon as *excessive sleepiness,* your rooting hunger late at night as *increased appetite*—and they prepare you for the cure by letting you know in what way you will feel better.

Just before I got my pills, Papakostas asked me how long it had been since I'd felt good for any appreciable time.

Good? I asked him.

"Symptom-free," he said, as if we had agreed that my feelings were symptoms.

"For how long?"

"Thirty days. Or more. At least a month."

I wanted to tell him that I was a writer, that I counted myself lucky to feel good from the beginning of a sentence to the period. I wanted to ask him whether he'd ever heard of betrayal, of disap-pointment, of mortality. Instead, I laughed—derisively, I suppose (was this the *irritability* of Item 10?)—and said I had no idea what a month of feeling good would feel like.

Of course, this only confirmed his diagnosis.

But *thirty days* is ringing in my ears as I head back to my car. I make a sudden decision: to duck into a restaurant, to order a glass of water with my meal, to start the trial not tomorrow morning but right now. I cannot resist the wish, the temptation: to lay down my pessimism at this altar, to put myself in the hands of these doctors, to take their investigational drug and let them cure me of myself. I gulp down my six golden pills.

BUT DRUGS DO WORK. By themselves, I mean, without the benefit of the placebo effect. Just ask the tuberculosis patients at Sea View Hospital in New York who, in 1952, took a derivative of hydrazine, a chemical that Germany used in the waning days of World War II to power its V-2s. The drug, called Marsilid, worked not only on their lungs but also on their heads; enough of them reported feeling euphoric—there was even a rumor that they were dancing in the wards—that doctors started to prescribe it for their melancholic patients.

In a society that's famously ambivalent about pleasure and the use of intoxicants to achieve it, however, it's not enough to take drugs to feel better. Especially for a drug company, it's better if you have an actual illness to treat, and a few years later, when it turned out that Marsilid prevented the brain from manufacturing an enzyme that broke down serotonin, an intriguing new chemical that had just been found in the brain, scientists had their disease. Depression, the new theory went, was not a psychological or existential condition, but a brain disease caused by a *serotonin deficiency* or some other *chemical imbalance*. Drug companies spread this gospel aggressively. In 1961, for example, Merck bought fifty thousand copies of *Recognizing the Depressed Patient,* a book by a doctor who had pioneered the serotonin theory and the use of drugs to treat it, to distribute to doctors who might not yet have heard that depression was the disease for which the new drugs were the cure.

But the evidence for the serotonin theory was circumstantial to begin with and has remained so for the last half century. Although scientists have mapped the jungle of nerve fibers through which serotonin makes its way from brain stem to synapse, analyzed the chemistry of that journey, and invented drugs that inhibit or encourage it along the trail, they have never proved that a serotonin deficiency actually exists in depressed people or, for that matter, figured out how much serotonin we ought to have in our brains in the first place. Nor have they explained certain inconvenient facts: for instance, that reserpine, a drug that *decreases* serotonin concentrations, also has antidepressant effects, or that so many people fail to respond to antidepressants—which, if they were really magic arrows aimed at a molecular bad guy, simply shouldn't be the case. Neither have they shown that in identifying the brain chemistry of a given mood or experience they have found the *cause,* rather than the *correlates,* the way the brain provides but doesn't originate that mood or experience. In the face of these dismal results, many scientists have begun to move on to theories about neurogenesis and cellular damage and other brain events of which serotonin may only be a marker, the finger pointing to the mood.

None of this stops doctors from continuing to manipulate serotonin in order to relieve depression. The omega-3s I am taking are thought to make neurons more supple and their outer membranes more receptive, allowing them to make the most efficient use of whatever serotonin is available. So far, however, the pills don't seem to be having an effect. Indeed, as I end my second week, I notice only one change. When I wake up early in the morning, when I crave my afternoon nap, when I find myself frustrated by my shortcomings or deflated by the seeming impossibility of getting done what I want to get done, when I read the newspaper and, like Ivan Karamazov with his catalog of atrocity, want to return my ticket, when I feel sorry for all of us, I wonder whether indeed I've been suffering from an illness all along. It is impossible to know, and I can't think of the experiment that will tell me.

But I can think of those thirty days. I'm not exactly obsessed with them, but I'm preoccupied with the idea that there are others right now in the midst of that month of resilience to setback and hardship, with the possibility that they are not simply luckier (or, as I think in my self-flattering moments, shallower), but *healthier* than I, that they have dodged a bullet that has caught me, that I can don some armor, make up prosthetically for what nature has, so these doctors say, denied me.

The third visit, the first one after I started the drugs, is shorter, more perfunctory than the first two. Papakostas moves briskly from one question to the next and looks at his watch if we digress. But the protocol calls for him to ask whether I have any questions. So I tell him I wasn't sure I'd understood him in our last meeting. How long was it that he thought I should be feeling good?

"For at least a month," he says.

I ask him why he wanted to know.

"People, when they're depressed," he says, "they get a sort of recall bias. They tend to feel that their past is *all* depressed."

Which means, I want to point out, that depression is more like an ideology than an illness, more false consciousness than disease.

This isn't the first impertinence I've stifled today. Earlier, he asked, "Are you content with the amount of happiness that you get doing things that you like or being with people who you like?"

"I'm not big on contentment," I said. *Is anyone?* I wondered. Is anyone ever convinced that his or her pursuit of happiness has reached its goal? And what would happen to the consumer economy if we began to believe that any amount of happiness is *enough*? "I'm sorry to seem dense," I explained, "but it's not how I usually think about things."

Papakostas was reassuring. "You know, this question condenses a lot of areas of life into just a number. It doesn't work well," he said. "Some questions we just don't like."

Well, if these are dumb questions, I wanted to shout, then *why are you asking them*? Why are we sitting here, turning me into an emotion McNugget? Why are we pretending that these answers mean anything? Indeed, if the whole point is to get at something that is in me but not of me, if I'm just the middleman here, the guy you've got to go through to get to the molecular essence of my troubles, then why ask *me* any questions at all?

Later, when he asked how many days I'd napped for more than thirty minutes in the last week and I told him four, he said, "See, some of the questions are really nice in terms of being objective," before putting me down for two points on that item.

"I suppose it would be easier if there were biochemical markers," I offered. "Otherwise, you're just stuck with language."

"Hey, we're psychiatrists," Papakostas said. "Language is good."

Now I was really confused. Hadn't we just spent the last half hour circumventing language's approximations? If language is good, then why wasn't he taping this visit, taking down my words instead of translating them into the tests' pale simulacrum of language, and from there into *just a number.* More to the point, if he said that the point was to reduce language to a number and then just a minute later said that language is good, why didn't his head explode?

For the same reason, I suppose, that he doesn't seem to think that consciousness itself, in all its insuperable indeterminacy, matters very much, which I discover when we meet two weeks later. I use my allotted question to ask Papakostas about a promising new experimental treatment for depression, one that uses an anesthetic drug called ketamine. A government psychiatrist was trying to bring ketamine in from the cold. In the psychiatric underground where drugs like LSD and psilocybin and ibogaine are used for transformative purposes, ketamine has a reputation for delivering a powerful and salutary, if terrifying, experience of being disembodied and dislocated, not unlike a near-death encounter. Papakostas is unfamiliar with the unofficial research, discredited since the sixties grew like an adipose layer over the therapeutic promise of psychedelic drugs, so I'm explaining the idea that with a single whack upside the head, one glimpse into the cosmos in all its glory and indifference can set you straight for a long time. I'm getting to the part about how inconvenient the economics of a one-time-only drug are for an industry addicted to one-a-days, when he interrupts.

"Sort of like ECT," he says, using the new, improved name for electric shock therapy. "The way it's supposed to reset your neurotransmitters. But we know that theory doesn't work, because ECT patients relapse."

"But isn't there a difference between ECT and ketamine?"

"Well, of course, ketamine works mostly on glutamate pathways . . ."

"No. I mean that you're conscious when you take ketamine and unconscious when you get ECT."

It's a distinction that seems lost on Papakostas, or maybe he just doesn't have time for a discussion of the nature of consciousness. Either way, you can't help but admire the purity of his devotion to the carnal, the way he's pared down psychic life to its bare bones. His is a spare and unrelenting pursuit, and his single-mindedness right now seems nearly ascetic.

Papakostas may be circumscribing my subjectivity in order to make it work for the drugs, but he's also renouncing his own, putting aside whatever curiosity he might have about the shape of the self, the objects of consciousness, the raw nature of our encounter, in order to make good his claim to possess the instruments of science. Armed with them, he can take my emotional measure and report my depression with the same dispassion and confidence as an astronomer telling the distance to a star. The truth thus derived, decontaminated of aspiration and expectation, is better, truer somehow, than the one we know through our credulous senses and fickle sensibilities. Maybe that's why I don't argue with him when he adds up my numbers and tells me that in the world behind the world, the one in which I am officially depressed, the survey says I'm getting better.

WHICH WAS NEWS TO ME. I hadn't been keeping track of my HAM-D and Q-LES-Q scores, but apparently my numbers were trending steadily toward health. I'm discomfited, disturbed, maybe even a little depressed at this, at my apparent inability to know my own inner state—not to mention the possibility that to get my thirty days, I will have to relinquish my own idea of happiness and settle for *symptom-free* living instead.

But come to think of it, maybe I am feeling a little better, and maybe it's not just that summer has hit its full stride or that I've had a couple of minor triumphs or that for the moment I'm successfully taking a stand against my own worst nature. Maybe the omega-3s are softening up my neurons. Maybe I'm happier in my meat.

Or maybe I'm identifying with the oppressor.

I arrive at my next visit resolved to get the dazzle out of my eyes and make my psychiatrist take account of the seams I think I'm seeing in the Matrix. But as I'm finishing up with the tests on my clipboard, a petite woman with short hair and large eyes comes into the waiting room. She's not quite looking at me as she quickly introduces herself, beckons me to follow her, and, before I can tell her that there must be some mistake, that I am Dr. Papakostas's patient, turns her back and briskly leads the way into the warren of offices beyond the waiting room.

He's gone away on vacation, I think. It is August, after all. But when we pass his office, there he is at his desk, leaning into his computer screen. He doesn't see me. I imagine that he has tired of my questions or that his colleagues have caught wind of our extra-curricular discussions, all that *language,* and have decided it's time to remind me who is asking the questions around here and pulled him off the case. But whatever the explanation, it is hard not to take this personally—which, of course, is exactly how a depressed person, whose disease makes him *rejection-sensitive,* would take it.

In fact, I can't seem to escape the gravitational field of my diagnosis today. When I tell the new psychiatrist I didn't catch her name, she repeats it carefully and slowly, as if to allow for my *psychomotor retardation.* When I explain why I am going to record our session (she asked, something Papakostas never did), she says,

"Oh . . . in-ter-est-ing," filling the spaces between syllables with professional smarm. She's running the numbers in her head, I think, wondering whether this will be the *difficult interview* that's worth three points on Item 8.

If the point of the switch was to make things more business-like, then Christina Dording was the perfect choice. She is cold and unflappable, her lines well rehearsed, her inflected concern perfectly pitched. She asks me whether I think my depression is a punishment for something that I've done, and I try joking: "It's an entertaining thought, but I haven't had that one." But she seems not to notice my attempt at humor. When I confess that I'm baffled, even after all these weeks, by the HAM-D's questions ("This past week, have you been feeling excessively self-critical?") that require me to parse words like *excessively* and *normally* and *especially* (something that Papakostas has dealt with affably by letting me ramble on until I say something that allows him to circle a number), she answers with crisp condescension: "If there's a comparator implied, it's always to when you're not depressed."

Her answer and its supercilious delivery make me wonder whether *I'm* the one asking silly questions. Maybe I'm the only person confused about whether *excessive* means more than I wish or more than I think others do or more than I think I ought to. Maybe her answer isn't as circular as it sounds. Maybe it means more than saying it's a problem when it's a problem and not when it's not. Maybe it isn't yet another denial of the basic assumption here—that they are the experts about my mental health, that depression isn't something I'm equipped to detect in myself because if I were, I'd be in the other study, the one for the Minor Depression I thought I had in the first place. Or maybe all these maybes, and my resulting inability just to blurt out a yes or a no, is just another example of my *excessive self-criticism.*

Dr. Dording and I are not off to a good start. Which makes it a little easier to interrupt the interview to ask her whether she really thinks self-criticism is pathological.

"Pathological?" she asks, as if she'd never heard the word. "I don't know if I'd call it pathological."

"Symptomatic, then," I offer.

"Well, it's certainly not optimal."

"Optimal," I say, deploying the therapist's repeat-and-pause tactic, hoping she'll tell me exactly how much self-criticism is optimal, and how she knows.

"*Certainly* not optimal." She does her own pause.

"But being self-critical is something that helps people achieve, isn't it?"

"Sometimes yes, sometimes no. I don't think being *excessively* self-critical is *ever* a great thing. No." She starts turning pages again and resumes the interview.

But I don't want to let it drop. I've come to pull back the curtain, and I go back to the first question I should have asked Papakostas. The numbers aside, I want to know, colleague to colleague, just between us pros, pinky promise I won't tell, do I really seem depressed to her? *Majorly* depressed? I try to get to the subject by asking her to tell me what she thinks the difference is between Major Depression and Dysthymia, a *DSM-IV* mood disorder that, if it has to be diagnosed, comes closest to capturing my melancholy.

"You're getting into close quarters here," she says.

In another world, one where psychiatrists *really* liked language, we might explore this slip, this unintended revelation of discomfort at my intrusion into her professional space—for she really means to say that I'm getting into fine diagnostic distinctions here. But she seems unaware of what she has just said as she explains, "Dysthymia is more low-level chronic. Minor Depression may or may not be long term, but it's typically less criteria than Major Depression."

And before I can ask her how any of this compares to what she actually *sees,* she closes the notebook and walks me out.

George Papakostas is a few paces in front of me as I round the corner of the reception desk. He's headed for the men's room. I decide to spare him strained pleasantries at adjoining urinals.

But I dawdle to the elevator, and he shows up just as it arrives. We ride down and walk out of the building together. I'm telling him how fascinating I'm finding this process, but how many questions I still have. I'm working up to asking him whether we can extend our next meeting somehow, maybe go out for lunch or something, so that I can debrief him. But he tells me he is going to Greece to visit his ailing father and that he won't be back in time. We shake hands good-bye.

I imagine that he is relieved to be done with me. I know how this looks to him, the patient challenging the boundaries of the professional relationship, the *What About Bob?* nightmare. Or I think I do. Maybe I don't know anything about this. Maybe what he really sees as we stand on the threshold of his concrete fortress is a conversation orchestrated by ion channels and neural pathways and axonal projections, two people deep in the grips of their chemicals, one of them still clinging (because of those chemicals, no doubt) to his old-fashioned idea that he's more than the sum of his electromolecular outputs, that a conversation like this one, not to mention recalcitrant unhappiness, might be complex and mysterious and meaningful.

I'M ALREADY DEFLATED when I arrive for my last interview. Of course, there's no place in the HAM-D to express this, to talk about the immeasurable loss that I think we are all suffering as science turns to scientism and bright and ambitious people devote their lives to erasing selfhood in order to cure it of its discontents. The HAM-D questions, Dording's grating solicitude, the banality of this exercise, the tyranny of the brain—they all seem as unassailable, solid, and impenetrable as the office building itself. I'm downright snarky when she asks me if I've been feeling guilty or self-critical.

"A constant feature of my life," I say. She ignores me.

But then she does something strange. She skips the Insight item, the one where she's supposed to ask whether I think I'm suffering from an illness and give me points if I don't think so. I ask her why.

"You typically don't ask," she replies. "It's atypical that a person is something other than a zero. Clearly psychotic people could have a two. There are occasions when you can get a one, like if a person thinks their lack of interest or energy doesn't have anything to do with being depressed. But, typically, people who are in here are a zero."

"So you would have to be either psychotic or believe that your symptoms are the result of some other conditions?"

"Yeah."

"As opposed to just saying, 'Well, you know, this is just how I am.'"

"That's a good question. I think that an answer like that would require an explanation; depending on the patient in the office, you need to talk a little more about an answer like that."

And I'm thinking that we should have this discussion, right now, because I am that patient, and I don't think I'm psychotic.

But that isn't going to happen. Instead, Dording is going to give me a physical. She goes to find out whether the exam room is available, returns to tell me that it is not and that I can wait or do it on my next visit.

"Next visit?" I ask. According to the protocol, this is my last.

"You're not coming in for the follow-up?" She looks as surprised as I am, as if no one would pass up that opportunity. I ask whether it would be any different from what we've been doing. It wouldn't, she says. So I tell her I'll skip the follow-up and wait for my exam.

Julie's also gone for vacation, so Caitlin takes my vitals and draws my blood. Then Dr. Dording comes in. She taps my knees, looks in my mouth, listens to my heart and lungs. When she asks me to follow her finger with my eyes, she leans in close and puts her hand on my bare knee. The touch of her fingertips is firm and cool and impersonal, my knee just a prop to hold her up.

She repeats her offer of a follow-up, then elaborates on something she mentioned at the end of our interview. "Give me one second here," she had said as she flipped the pages of my binder.

"Look at your scores. Nice response." Now she says, in case I didn't get it the first time, "I think you've done very well, you're *much improved*." She doesn't ask whether I agree or explain why, if I'm better, I'd need *follow-up,* why I need to do more than buy some fish oil at the Whole Foods next door.

If, that is, I'd been taking fish oil for the last eight weeks.

I ask her whether I was on the placebo or the drug. She's befuddled for a moment. "I don't think we unblind the study," she says. She deliberates over my paperwork. "No, not in this one. No unblinding."

I protest, "I don't get to find out?"

It's as if she's never been asked, as if no one in the whole history of clinical trials had ever wanted to know which side they'd been a witness for.

"No," she says. "But you had a good response." She's chipper now, as if she's trying to convince me that I ought to take my improvement and go home happy, another satisfied customer. And really it doesn't matter. Because the point here is not to teach me anything about myself or for them to learn anything from me. It's not even to prove whether omega-3s work. It's to strengthen the idea that this is what we are: machines fueled by neurotransmitters at the mercy of our own renegade molecules.

ONCE UPON A TIME, the scientific explanation for depression sounded something like this:

> If one listens patiently to the many and various self-accusations of the melancholiac, one cannot in the end avoid the impression that often the most violent of them are hardly at all applicable to the patient himself, but that with insignificant modifications they do fit someone else, some person whom the patient loves, has loved or ought to love. . . . So we get the key to the clinical picture—by perceiving that the self-reproaches are reproaches against a loved object which has been shifted onto the patient's own ego.

For a modernist like Freud, who wrote *Mourning and Melancholia* in 1917, depression was embedded in history, personal and cultural, and untangling that history, rescuing it from the oblivion of the unconscious by turning it into a coherent story, was the key to a cure. This fascinating and tragic notion, that we carry within us an other whom we can never fully know but whom we must try to, has been carted to the dustbin of history by the Dordings and the Papakostases of the world. The extravagant hermeneutics, the because-I-said-so epistemology, the unfalsifiable claims of psychoanalysis—not to mention its sheer inefficiency, its indifference to the possibility that analysis might be interminable—have given way to inventories and brain scans and double-blind studies. Freud's unconscious, the repository of that which is too much to bear and which will only stop tormenting us to the extent that we give it language, has been replaced with an unconscious populated by carbon and hydrogen and nitrogen and oxygen, the basic building blocks of the material world, essential but forever dumb.

Still, I'm not exactly pining for Freud as I leave Mass General for the last time. He got too much wrong, some of it inexcusably so. Indeed, as I dodge the lunch-hour scurry on the hospital zone streets, the doctors in blue scrubs hurrying between buildings, wan patients wheeling IV stands down the sidewalks, ambulances and private cars delivering a legion of the sick to this city of hope shimmering in the late summer heat, to lay their time and money and dignity down upon the altar of science, I am once again struck by the soreness of my temptation to do the same, to believe that I am indeed better now, that the person who drives down Storrow Drive having these thoughts, passing other I's having their own thoughts, all of us convinced that we are inside ourselves just as surely as we are inside our cars—that I am wrong about who I am, that we are all wrong. That scientists peering into the darkness in our skulls will eventually illuminate it entirely and show us that such thoughts and the conviction with which they are held are only accidental: spandrels of our cerebral architecture that can be rearranged with surgical precision. And just as we once were play-things for the gods or sinners poised over a fiery pit or enlightened rationalists cogitating our way to the truth of ourselves, we will become the people who needn't take ourselves too seriously, who will cease to mistake the vicissitudes of personal history for the vagaries of personal biochemistry, who will give up the ghost for the machine.

Because irresistible ideas about who we are only come along every so often. And here at Mass General they're working with a big one. They have figured out how to use the gigantic apparatus of modern medicine to restore our hope: by unburdening us of self-contradiction and uncertainty, by replacing pessimism with optimization, by inventing us as the people who seek Life Enjoyment and Satisfaction and forge ahead, who buy from the pharmacy what we need to forge ahead toward Well Being unhindered by Depressive Symptomatology, to pursue antidepression if not happiness. Who can resist this idea that our unhappiness is a deficiency that is in us but not of us, that is visited upon us by dumb luck, and that can be sent packing with a dab of lubricant applied to a cell membrane?

The epiphany makes me wonder if I've been churlish to Christina Dording; maybe I should take her word for it, accept that I am better now, and thank her. But remorse lasts only as long as it takes to get the results from the lab to which, out of curiosity, I sent a sample of my pills. There wasn't a drop of fish oil in them; I was on the placebo.

3

SEXUAL ORIENTATION: GAY SCIENCE

W HEN HE LEAVES HIS TIDY APARTMENT in an ocean-side city somewhere in America, Aaron turns on the radio to a light rock station. "For the cat," he explains, "so she won't get lonely." He's short and balding and dressed mostly in black, and right before I switch on the recorder, he asks me for the millionth time to guarantee that I won't reveal his name or anything else that might identify him. "I don't want to be a target for gay activists," he says as we head out into the misty day. "Harassment like that I just don't need."

Aaron sets a much brisker pace down the boardwalk than you would expect of a doughy fifty-one-year-old, and once reassured of anonymity, he turns out to be voluble. Over the crash of the waves, he spares no details as he describes how much he hated the fact that he was gay and how the last thing in the world he wanted to do was to act on his desire to have sex with another man.

He managed to maintain his celibacy through adolescence, which included a four-year stint at a college with the usual loose sexual mores, and into adulthood. But when, in the late 1980s, he found himself so "insanely jealous" of his roommate's girlfriend that he had to move out of their apartment, he knew the time had come to do something. One of the few people who knew that Aaron was gay showed him an article in *Newsweek* about a group offering "reparative therapy"—psychological treatment for people who wanted to become "ex-gay."

"It turns out that I didn't have the faintest idea what love was," he says. That's not all he didn't know. He also didn't know that his same-sex attraction, far from being inborn and inescapable, was a thirst for the love that he had not received from his father, a cold and distant man prone to angry outbursts, coupled with a fear of women kindled by his intrusive and overbearing mother, all of which added up to a man who wanted to have sex with other men just so he could get some male attention. He didn't understand any of this, he tells me, until he found a reparative therapist with whom he consulted by phone for nearly ten years, attended weekend workshops, and learned how to "be a man."

Aaron interrupts himself to eye a woman jogging by in shorts. "Sometimes there are very good-looking women at this boardwalk," he says. "Especially when they're not bundled up or anything like that." He remembers when he started noticing women's bodies, a few years into his therapy. "The first thing I noticed was their legs. The curve of their legs."

He's dated women, had sex with them even, although "I was pretty awkward," he says. "It just didn't work." Aaron has a theory about this: "I never used my body in a sexual way. I think the men who actually act it out have a greater success in terms of being sexual with women than the men who didn't act it out."

Not surprisingly, he's never had a long-term relationship, and he's pessimistic about his prospects. "I can't make that jump from having this attraction to doing something about it." But, he adds, it's wrong to think that "if you don't make it with a woman, then you haven't changed." The important thing is that "now I like myself. I'm not emotionally shut down. I'm comfortable in my own body. I don't have to be drawn to men anymore. I'm content at this point to lead an asexual life, which is what I've done for most of my life anyway. I'm a very detached person."

It's raining a little now. We stop walking so I can tuck the microphone under the flap of Aaron's shirt pocket, and I feel him recoil as I fiddle with his button. I'm remembering his little cubicle of an apartment, its unlived-in feel, and thinking that he may be the sort of guy who just doesn't like anyone getting too close; but it's also possible that therapy has caused him to submerge his desire so deep that he's lost his motive for intimacy.

That's the usual interpretation of reparative therapy—that, to the extent that it does anything, it leads people to repress, rather than change, their natural inclinations; that its claims to change sexual orientation are an outright fraud perpetrated by the religious right on people who have internalized the homophobia of American society and personalized the political in such a way as to reject their own sexuality and stunt their love lives. But Aaron scoffs at these notions, insisting that his wish not to be gay wasn't a religious thing: he's a nonobservant Jew who is disturbed by the strong influence of the religious right in reparative therapy circles. It wasn't about politics, either. He's a lifelong Democrat who volunteered for George McGovern, has a career as a public servant, and thinks George Bush is a war criminal. It wasn't a matter of ignorance: he has an advanced degree. And it *really* wasn't a psychopathological thing—he scoffs at the idea that he's suffering from internalized homophobia. He just didn't want to be gay, and, like millions of Americans who are dissatisfied with their lives, he sought professional help and reinvented himself.

Self-reconstruction is what people in my profession specialize in, but when it comes to someone like Aaron, we draw the line. All the major psychotherapy guilds have barred their members from researching or practicing reparative therapy on the grounds that it is inherently unethical to treat something that is not a disease, that it contributes to oppression by re-pathologizing homosexuality, and that it is dangerous to patients whose self-esteem can only suffer when they try to change something about themselves that they can't (and shouldn't have to) change.

As a result, reparative therapy has become an outpost of the religious right, an association that bothers Aaron, especially when he has to prove that he's not pulling a Ted Haggard. But he knows why he has to do all this explaining. He threatens a political and scientific consensus that has been emerging over the last century and a half: that sexual orientation is inborn and immutable, that efforts to change it are bound to fail, and that discrimination against gay people is therefore unjust.

Reasoning like this has been crucial to the struggle for gay rights, and it's become, outside of the religious right, a truism so mainstream that Subaru ran an ad campaign featuring the slogan: "It's not a choice, it's the way we're built." (And earned itself the nickname "Lesbaru" and perhaps a market niche in the process.) But if Aaron isn't a poseur motivated by an antigay agenda, then he's walking evidence that the consensus is wrong, that this justification for gay rights may not be as sound as we might wish. While scientists have found some intriguing biological differences between gay and straight people, the evidence so far stops well short of proving that we are born with a sexual orientation that we will have for life. Even more important, some research shows that sexual orientation is more fluid than we have come to think, that people, especially women, can and do move across customary sexual orientation boundaries, and that there are ex-straights as well as ex-gays. Much of this research has stayed below the radar of the culture warriors, but the reparative therapists, who would like very much to convince as many gay people as possible to seek professional help, are watching it closely and preparing to use it to enter the scientific mainstream and advocate for what they call the right of self-determination in matters of sexual orientation. If they are successful, gay activists, who have built public support for gay rights on the belief that homosexuality is inborn and immutable, may soon find themselves scrambling to make sense of a new scientific and political landscape.

IN 1838, A TWENTY-YEAR-OLD Hungarian man killed himself and left a suicide note for Karl Benkert, a fourteen-year-old bookseller's apprentice in Budapest whom he had befriended. In it, the suicide explained that he had been cleaned out by a blackmailer who was now threatening to expose his homosexuality and that he couldn't face either the shame or the potential legal trouble that would follow. Benkert, who eventually moved to Vienna and changed his name to Karoly Maria Kertbeny (apparently, a Hungarian name had more cachet in Viennese salons), later said that the tragedy left him with "an instinctive drive to take issue with every injustice." And in 1869, a particularly resonant injustice occurred: a penal code proposed for the Prussian nation, which was created when Bismarck annexed the Austrian states, included an antisodomy law much like the one that had given his friend's extortionist his leverage. Kertbeny published a pamphlet in protest, writing that the state's attempt to control consensual sex between men was a violation of the fundamental rights of man.

Nature, he argued, had divided the human race into four sexual types: *monosexuals,* who masturbated; *heterogenits,* who had sex with animals; *heterosexuals,* who coupled with the opposite sex; and *homosexuals,* who preferred people of the same sex.

Kertbeny couldn't have known that of all his literary output, these latter two words would be his only lasting legacy. But in fact, although homosexuality had occurred throughout history, up to that point no one had claimed that homosexuals constituted a category of human being, a sexual subspecies that differed from other human populations in some essential way.

Kertbeny wasn't alone in creating a sexual taxonomy. Another antisodomy-law opponent, lawyer Karl Heinrich Ulrichs, proposed that homosexual men, or *Uranians,* as he called them (and he openly considered himself a Uranian, whereas Kertbeny was coy about his preferences), were actually a third sex, their attraction to other men a manifestation of the female soul residing in their male bodies. Whatever the theoretical differences between Ulrichs and Kertbeny, they agreed on one crucial point: that sexual behavior was the expression of an identity into which we were born, a natural variation of Homo sapiens. In keeping with the post-Enlightenment notion that we are morally culpable only for what we are free to choose, homosexuals were not to be condemned or restricted by the state. Indeed, this was Kertbeny's and Ulrichs's purpose: sexual orientation, as we have come to call this biological essence, was invented in order to secure freedom for gay people.

But replacing morality with biology, and the scrutiny of church and state with the observations of science, invited a different kind of condemnation. By the end of the nineteenth century, homosexuality was increasingly the province of psychiatrists such as Magnus Hirschfeld, a gay Jewish Berliner. Hirschfeld was an outspoken opponent of antisodomy laws and championed tolerance of gay people, but he also believed that homosexuality was a pathological state, a congenital deformity of the brain that may have been the result of a parental "degeneracy" that nature intended to eliminate by making the defective population unlikely to reproduce. Even Sigmund Freud, who thought people were polymorphously perverse by nature and urged tolerance for homosexuality, thought that heterosexuality was essential to maturity and psychological health.

Freud was pessimistic about the prospects for treatment of homosexuals, but doctors abhor an illness without a cure, and the twentieth century saw therapists inflict on gay people the best of modern psychiatric practice, which included, in addition to interminable psychoanalysis and unproven medications, treatments that used electric shock to associate pain with same-sex attraction. These therapies were largely unsuccessful, and, particularly after the Stonewall Riots of 1969 (the event that historians credit with initiating the modern gay rights movement), patients and psychiatrists alike started to question whether homosexuality should be considered a mental illness at all. Gay activists, some of them psychiatrists, disrupted the annual meeting of the American Psychiatric Association (APA) for three years in a row, until in 1973 a deal was brokered: the APA would delete homosexuality from its *Diagnostic and Statistical Manual of Mental Disorders (DSM)* immediately, and, furthermore, it would add a new disease: Sexual Orientation Disorder, in which a patient can't accept his or her sexual identity. The culprit in SOD was an oppressive society, and the cure for SOD was to help the gay patient overcome oppression and accept who he or she really was. (SOD has since been removed from the *DSM.*)

The APA cited various scientific papers in making its decision, but most of them relied on circular reasoning: if you remove homosexuality from tests of psychopathology, then homosexuals as a group no longer test as pathological. Some members protested that this transparently political move was a dangerous corruption of science. As one dissenting psychiatrist complained, "If groups of people march and raise enough hell, they can change anything in time. Will schizophrenia be next?" And the protesting members' impression was confirmed when the final decision was made not in a laboratory but at the ballot box, where the membership voted to authorize the APA to delete the diagnosis. It may be the first time in history that a disease was eliminated by the stroke of a pen. It was certainly the first time that psychiatrists determined that the cause of a mental illness was an intolerant society. And it was a crucial moment for gay people, at once getting the psychiatrists out of their bedrooms and giving the weight of science to Kertbeny's and Ulrichs's claim that homosexuality was an identity, like race or national origin, that deserved protection.

THREE DECADES AFTER THE DELETION, some groups are still marching and raising hell about the mental health industry's role in the struggle for gay rights. Chief among them is the National Association for the Research and Therapy of Homosexuality (NARTH), an organization founded by Charles Socarides, a psychiatrist who led the opposition to the 1973 APA vote. "They will wipe the floor with us," Socarides said when he launched NARTH, "but we will wear our wounds as badges of courage."

When I attended the group's national conference in Orlando in November 2006, I was treated to some of this firebrand rhetoric. Joe and Marian Allen, for instance, took to the lectern to tell us how God had called them to "testify" about their gay son who was murdered by his lover, a tragedy that they managed to twist into a cautionary tale about what happens when a "struggler" is told by a "well-meaning therapist" that he was "born gay" and can't change it. And Kermit Rainman, an ex-gay Focus on the Family minion, cheerfully explained the gay agenda to me. "It's doing whatever you want, whenever you want, with whoever you want, wherever you want."

"Well, just for the sake of argument," I said, "what's wrong with that?"

"I'm sure the people who follow that agenda believe what they believe, but they don't realize that they're pawns in a great cosmic battle, that they are perpetrating a lie."

"Pawns of . . . ?"

"Satan," he informs me, "is the author of lies, chaos, and confusion."

But hellfire and brimstone don't rain down on this conference nearly as much as I expected. In fact, it's disappointingly similar to every other convention: bad coffee, worse Danish, stale hotel air, dry-as-dust lectures. Dean Byrd, a psychologist and a professor at University of Utah Medical School and a long-time NARTH leader, methodically lays out his case that sexual orientation is malleable.

If NARTH's strategy is to seek a place at the table by behaving like serious scientists, Byrd's modulated approach, tedious as it may be, is just what the doctor ordered. Sometimes he's puckish, as when he says, "When it comes to homosexuality, I'm pro-choice," a comment that's sure to get a rise out of a crowd well versed in the other moral disaster of 1973, and sometimes he's glib ("The proper answer to the nature-nurture question is yes"). Mostly, however, he's just workmanlike as he reviews the research—much of it, he is delighted to point out, conducted by the "activists themselves." He cites papers from Denmark, the first place that legalized civil unions and perhaps, he says, the most gay-friendly place in the world. In that country, according to the studies, gay people turned out to have mental illness at a higher rate than straights, which proves, he says, that an intolerant society is not the culprit when gay people suffer.

Byrd also describes the studies that show that the identical twin brother of a gay man has *only* a 50 percent chance of being gay himself, which may be twice the rate among fraternal twins, but still, he argues, far from the 100 percent you would expect if sexual orientation is purely genetic. And he recounts his own work with gay people, even shows us a video of one of his treatment sessions (which is just as dull and uneventful as any other videotaped treatment session), and gives a plausible-sounding assessment of the prospects for patients of reparative therapists: that one-third of them will become heterosexual, one-third will remain gay, and one-third will move a few notches along the Kinsey scale, enough to leave the lifestyle and limit their unwanted feelings and behavior.

Like everyone else here, Byrd is very excited about an article that appeared in *Archives of Sexual Behavior* in 2003. It was a small study, only two hundred subjects, but it concluded that gay people could indeed change their sexual orientation, that the change was not merely religiously motivated repression or politically motivated bluster but rather some fundamental shift in desire; the researcher concluded that even the people who showed little benefit from reparative therapy didn't seem to derive harm, and that much more research needed to be done. The study was full of caveats and received withering criticism from other scientists who claimed that it relied on a skewed sample—mostly people handpicked by therapists such as Byrd and NARTH president Joseph Nicolosi—but it had passed peer review, and, even more important, it had been conducted by none other than Robert Spitzer, the psychiatrist who had brokered the deal that deleted homosexuality from the *DSM*.

Spitzer also called for an end to the ban on research into reparative therapy, and one psychologist who has taken him up on his call is being welcomed in Orlando like a conquering hero. Elan Karten is an unassuming young man who wears a yarmulke and recently got a doctorate from Fordham after writing a dissertation on ex-gay men. Karten got the go-ahead for his study only by positioning it as an inquiry into the type of people who seek reparative therapy rather than as an exploration of its efficacy. He did manage to sneak in some of that research as well and reached conclusions similar to Spitzer's. So far, however, Karten's work hasn't joined Spitzer's in the academic press; peer reviewers, objecting that the study revives the notion that homosexuality is a mental illness, have prevented Karten, an academic unknown, from publishing his work.

I tell Aaron about some of what I witnessed at NARTH: the time that Nicolosi, recalling one of his antagonists at a recent American Psychological Association convention, said, "I knew that she was a lesbian. I don't know why, maybe because she was wearing a muscle shirt"; the way that Byrd trailed off into a world-weary sigh after he granted that people have the right to stay gay if they want; the casual homophobia I heard in the halls. But Aaron has attended a NARTH meeting. He already knows that its members and leaders don't like gay people much. And he's much more concerned with a different kind of intolerance, the kind that Karten and Spitzer have encountered, and he sees the irony in his situation. "Not all homosexual men want to lead a gay lifestyle. They [gay activists] shouldn't be threatened by that. I mean, here I am, as a liberal, telling gay people to accept diversity."

WHEN HE TALKS ABOUT DIVERSITY, Aaron probably isn't thinking about what goes on inside a low-slung cinderblock building under a freeway ramp hard by piers and cranes of Seattle's shipping port. The sign on the door says Sex Positive Community Center, but everyone here calls it the Wet Spot. It's Thursday night, time for the Grind, the weekly dance. The music is already pounding, but before I can even give the doorwoman my ten-dollar entrance fee, she is insisting that I read the rules—obtain consent for everything, mind your bodily fluids, no drugs or alcohol, and what happens at the Wet Spot stays at the Wet Spot—and assent to them in writing before I enter. And you don't say no to a broad-shouldered woman in full dominatrix regalia. (I did get a dispensation to report on the scene, as long as everyone stayed anonymous and I took no pictures.)

I've already been beyond the partition that screens the dance floor from curious eyes on the street. A couple of nights ago, I attended the mummification workshop and watched as a man snapped on a pair of blue latex gloves (hygiene is very important at the Wet Spot), took out an industrial-size roll of Saran, wrapped it around a very large and naked volunteer until she was bound to the table like a giant bird under cellophane in a grocery store display case, and demonstrated the sexual frolic that could then ensue.

Nobody bothered getting dressed up for the workshop or for the drop-in social the next day. In fact, most of the drop-ins were pale and flabby and dressed in polyester, the ambience more Star Trek convention than Plato's Retreat. The conversation passing around the table along with the M&Ms and Doritos was clinical, a little geeky even, until I brought up the subject of biology. They all wanted to set me straight about this: the sex researchers have it wrong about why they are the way they are.

"Look, I don't think I was made this way," a gray-haired man in his early fifties said by way of explaining his interest in bondage. "It's something I found and I turned out to like it. I didn't go out of my way to look for it, and it was plenty weird at first, but I was intrigued, and now it's a big part of my life."

"You'll see," said a woman. "Come to the Grind and see what people do when they start playing with their sexuality." And, she added, when there's some "eye candy" around.

So here they are in their Thursday best: women in bone-and-lace bustiers, men in leather pants and Utilikilts, canvas skirts with grommeted pockets in their pleats. Or in nothing at all.

And they're not just lecture-demonstrating. Couples and trios kiss and squeeze, changing partners as if attending a square dance hosted by the Marquis de Sade.

But what's really got my attention, enough to make me ignore the pretty young couple canoodling on the mattress behind me, is in the other half of the room: torture racks, some sections of chain-link fence, and shiny metal loops dangling from the ceiling. And people are using them. But, at least to judge from the rapture on their faces, the violence, like all the debauchery, is also delicious to the participants. I feel as if I've stumbled into Hieronymus Bosch's *The Garden of Earthly Delights,* the center panel, the one flanked by Paradise and Hell.

Kermit Rainman would probably appreciate this comparison, although he'd remind you that Bosch wasn't exactly on the side of the sybarites. It's pretty clear that this is his satanic dystopia, and if you don't want to find yourself on Rainman's side, you have to stay focused on that strange tenderness, on the fact that everyone here is trying to make it possible for you to do whatever you want, whenever you want, with whomever you want.

You also have to let go of familiar ideas about equality and violence and gender, but, above all else, of sexual orientation. At least, that's the view of Michael, one of the founders of the Wet Spot, who won't let me use his real name. He's a droll sixty-four-year-old grandfather with a trim beard and half-glasses. (His five-year-old grandson is visiting this week; he accompanies Michael to our meetings at various restaurants around the city.) Michael doesn't think that sexual orientation can possibly be immutable, and that's not only because he's a molecular biologist who once specialized in genetics and knows a thing or two about mutation. It's also because he's dedicated himself to expanding his sexual horizons, and he's found himself doing things he never imagined he would—not just whipping people and enjoying it, but also having sex with men.

"It's an acquired taste," he tells me, "like sushi. You know, you hear 'raw fish,' and you think 'no way,' and then someone comes up and says, 'Try a bite of this,' and then you say, 'Oh, this is not so bad.'" Which doesn't mean, Michael tells me quickly, that he is gay or even bi. "I didn't say, 'Oh my God, look what I've been missing,' and leave my wife. I just thought, 'This is nice. This is another thing in life that's nice.'" In fact, since he's incorporated men into his sex life—most of these encounters involve his wife, to whom he's been married for more than forty years, and with whom he shares a love life based, he says, on honesty rather than on monogamy—he's come to think that the straight/gay/bi taxonomy is wrongheaded. "My binary is mono/poly," he says. "You either stick to one thing or you're fluid."

Hang out at the Wet Spot for a week, and you'll hear this over and over: that the categories of sexual orientation don't do justice to the range of human sexual expression, that the real problem is that too many people are "monosexuals."

To be pigeonholed into a fixed sexual orientation, they tell me, is politics at its most personal, a form of oppression that, by forcing people to categorize themselves, restricts their self-expression, their access to pleasure and intimacy. "I used to identify myself as lesbian," a woman told me. "But all of a sudden I realized that the more I say, 'Well, that doesn't attract me,' the fewer opportunities I have. So, I just kind of open myself up to being just in the moment, what's hot right now."

So does this mean that sexual orientation is a choice? When I ask this of a group of Wet Spotters gathered at an upscale bistro for a weekly cocktail hour, the discussion is animated. They search for ways to explain that sexual exploration, trying new pleasures, and maybe even finding compelling what once was repellent or discovering that their preferences have changed—none of this is a simple matter of choice. There's a lot of good-natured bickering about the right analogy, and some not-so-good-natured suspicion directed at me, at the possibility that letting me in on this conversation will play into the hands of the religious right.

Finally, a man who has been quiet speaks up. "The course of a river is a function of hydrology and geology," he says. "It changes over time, it changes with upheaval, it responds to conditions. But you can't force it to run the way you want; you can't choose where it is going to go. That's the problem with the religious right saying it's a choice. But the problem with the categories, with the whole gay/straight/bi thing, is how hard they make it to just ride the river."

TWO RUMORS CRACKLE THE AIR at the NARTH conference: the one purporting to explain why the main speaker didn't show up (because he thinks that NARTH hasn't sufficiently disavowed the statement of one of its board members that American slavery might have been an overall good thing for the Africans) and the one about the protest that may or may not breach the doors at high noon on Saturday. That one is confirmed when we are instructed not to respond to the protestors ("Sing a hymn or pray instead") who are gathering on the other side of the small pond in front of the hotel. You have to stand in the entrance and crane your neck around some low bushes to see them as they put on their duck outfits, hoist their signs ("Stop Ducking the Truth"; "NARTH is Goofy"), make quacking noises, and yell "Shame!" in our general direction. Dean Byrd looks out the door, shakes his head, and laughs with the rest of the small crowd when a man behind him says, "Quack, quack? *They're* the queer ducks."

I wait until the foyer is empty before I head out into the Florida sun to see the protest up close. Wayne Besen, a tall man in a polo shirt, is pulling the props and the costumes out of his car trunk. He runs Truth Wins Out, an organization devoted to debunking the research of the ex-gay movement. He minces no words about Spitzer's research: "One of the most poorly constructed studies in the history of science, a travesty." And he calls reparative therapy "intelligent design for gay people." Besen thinks the stakes of the scientific battle are impossible to overstate. "Americans are not cruel. If they think that being gay is inborn and can't be changed, they are going to be very sympathetic to full equality for gay people," he says. "We win this argument, the gay rights struggle would be done."

Besen is sure that science is on the verge of giving gay people their slam dunk. After all, he says, study after study shows that homosexuality is biological in origin. In the last fifteen years, researchers have discovered differences in brain anatomy between gay and straight men and found that the 6 percent of rams that have sex exclusively with other rams (just one example of more than three hundred species in which homosexual behavior has been observed) have a similar neuroanatomical difference. Scientists have also identified a gene sequence on the X chromosome that is common to many gay men; have traced genealogies to show that homosexuality runs in families, on the maternal side; have proved that a man's likelihood of being gay increases with the number of older brothers he has, which scientists attribute to changes in intrauterine chemistry; and have learned how to use magnetic resonance imaging to detect sexual orientation by watching the brain's response to pornography. Findings in the field of anthropometrics have yielded intriguing results: gay men's index fingers, for instance, are more likely than straight men's to be equal in length to their ring fingers; and gay men have larger penises than straight men. These findings all seem to support Besen's contention that being gay is essentially biological and should remain beyond the reach of either law, morals, or medicine.

But Besen hasn't been to the Wet Spot. And as attractive as this line of reasoning is, gay activists don't unanimously endorse it. "One thing I find troubling within the gay community is a lot of people feel if they can make that claim strongly enough, that's going to give them equal rights," Sean Cahill, the director of the Gay and Lesbian Task Force Policy Institute, told me. "But I don't think it really matters," he said, pointing out that believing that sexual orientation is biological doesn't cause people to support gay rights. Indeed, many social scientists think that the beliefs are merely correlated, that people who hold one tend to hold the other.

Gay rights lawyers also downplay the importance of biological accounts of sexual orientation. The Supreme Court has ruled that the immutability of a group's identifying characteristics is one of the criteria that entitle it to heightened protection from discrimination (and some of the early cases establishing gay rights were decided in part on those grounds), but, according to Suzanne Goldberg, the director of the Sexuality and Gender Law Clinic at Columbia University, there is a far more fundamental reason for courts to protect gay people. "Sexual orientation does not bear on a person's ability to contribute to society," she told me. "We don't need the science to make that point."

Jon Davidson, the legal director of Lambda Legal, agreed, adding that if courts are going to ask about immutability, they shouldn't focus on biology. Instead, they should focus on how sexual orientation is so deeply woven into a person's identity that it is inseparable from who they are. "What will it mean to say to somebody that the only way to get equal treatment is if you change what's really a fundamental part of your identity—how you relate to other people in the most intimate parts of your life?" In this respect, Davidson said, sexual orientation is like another core aspect of identity that is clearly not biological in origin: religion. "It doesn't matter whether you were born that way, it came later, or you chose," he said. "We don't think it's okay to discriminate against people based on their religion. We think people have a right to believe whatever they want. So why do we think that about religion but not about who we love?"

Cahill, who said he doesn't think he was born gay, also pointed out that even if it is crucial for public support, essentialism has a dark side: the re-medicalization of homosexuality, this time as a biological condition that can be treated, perhaps even in utero. Michael Bailey, a Northwestern University psychologist who has conducted some of the crucial studies of the genetics and neurochemistry of sexual orientation, infuriated the gay and lesbian community with a paper arguing that should prenatal markers of homosexuality be identified, parents ought to have the right to abort the potentially gay fetuses. "It's reminiscent of eugenicist theories," Cahill told me. "If it's seen as an undesirable trait, it could lead in some creepy directions." These could include not only abortion but gene therapy or even modulating uterine hormone levels to prevent the birth of a gay child.

Lisa Diamond, a professor of psychology at the University of Utah, may have the best scientific reason of all for activists to shy away from arguing that homosexuality is inborn and immutable: it's not exactly true. She doesn't dispute the findings that show a biological role in sexual orientation, but she thinks far too much is made of them.

"The notion that if something is biological, it is fixed—no biologist on the planet would make that sort of assumption," she told me from her office at the University of Utah. Not only that, she said, but the research—which she pointed out is conducted almost exclusively on men—hinges on a very narrow definition of sexual orientation. "It's what makes your dick go up. I think most women would disagree with that definition," she said, not only because it obviously excludes them but because sexual orientation is much more complex than that. That complexity is exactly what is lacking in most research, which tends to focus on the observable aspects of sexuality. "An erection is an erection," she said, "but we have almost no information about what is actually going on in terms of the subjective experience of desire."

Diamond has spent the last twelve years doing her part to fill in this gap by following a group of seventy-nine women who originally described themselves as nonheterosexual, and she's found that sexual orientation is much more fluid than activists such as Besen believe. "Contrary to this notion that gay people struggle with their identity in childhood and early adolescence, then come out and ride off into the sunset," she said, "the more time goes on, the more variability comes out. Women change their identities and find their attractions changing." In the first year of her study, 43 percent of her subjects identified themselves as lesbian, 30 percent as bisexual, and 27 percent as unlabeled. By year ten, those percentages had changed significantly: 30 percent said they were lesbian, 29 percent said they were bisexual, 22 percent wouldn't label themselves, and 7 percent said they were now straight. Across the entire group, Diamond found that only 58 percent of her subjects' sexual contacts were with women; in year eight, even the women who identified as lesbians reported that between 10 and 20 percent of their sexual contacts were with men. From her data, Diamond concluded that the categorization of women into gay, straight, and bisexual misses an important fact: that they move back and forth among these categories, and that the fluidity that allows them to do so is as crucial a variable in sexual development as their orientation.

Like the Seattle sex-positives, Diamond cautioned that it's important not to confuse plasticity, the capacity for sexual orientation to change, with choice: the ability to change it at will. "Trying to change your attractions doesn't work very well, but you can change the structure of your social life, and that might lead to changes in the feelings you experience." This is a time-honored way of handling unwanted sexual feelings, she pointed out. "Jane Austen made a career out of this: people fall in love with a person of the wrong social class. What do you do? You get yourself out of those situations."

For the women in Diamond's study who tell her, "I hate straight society, I don't want to be straight," Austen's solution is an effective treatment for unwanted other-sex attraction. "If you're around women all the time and you are never around men, you are probably going to be more attracted to women," she said. Such women sometimes end up falling in love with women, and their sexual feelings follow. And it can work the other way, Diamond said: women who identify themselves as gay or bisexual sometimes find themselves, to their own surprise, in love with men with whom they then become sexual partners. Indeed, she said, "Love has no sexual orientation."

Diamond's mentor at Cornell University, Daryl Bem, thinks that Diamond's subjects' ability to "start with love" indicates that "sexual orientation is primarily important as a political concept. It wouldn't be salient if we lived in a society that didn't care very much." After all, he said, "You fall in love with a particular person, not with a gender," so in a society where sexual orientation is less salient, people would have more choices, more opportunities to love. "We would use other criteria besides the sex of the partner to guide our attractions."

Bem has some experience in these matters. Although he is gay, he was happily married for twenty-nine years to a woman with whom he had two children (and, he said, "great sex"). "If I had been born thirty years later," said Bem (he was born in 1931), "I might have turned out just gay with a gay partner. But I was dating women, and as it happened, I fell in love with one of them."

Bem doesn't think that his marriage changed his basic orientation—he's never fallen in love with another woman, and he is currently living with a male partner—but he sees no reason why people shouldn't be allowed to try to change their sexuality. "I'm losing my hair. I want a hair transplant, and people say, 'Why do you want to change yourself? You're just giving in to society's prejudice against bald men.' But would you take steps to deny me a hair transplant because I'm just a victim of society's prejudices? I have the same feelings about someone who doesn't want to be gay." For Bem (who said he never got the transplants but did wear a toupee for five years), the analogy indicates how unimportant the whole notion of sexual orientation is, or ought to be. Indeed, he's a walking example of what would happen if our ideas about sexual orientation didn't control our love lives—we'd all have more choices about whom to love.

"Maybe men would still prefer blondes," he said, "but they wouldn't care what sex they were."

As confusing as this sexual disorientation might be, it would at least help to reestablish the boundary between science and politics when it comes to human sexuality. As Diamond pointed out, the search for the biological origins of sexual orientation is inescapably political and thus a misuse of science. (Bem, seconding this insight, pointed out that "you never hear people asking where heterosexuality came from.") "We live in a culture where people disagree vehemently about whether sexual minorities deserve equal rights," Diamond told me. "People cling to this idea that science can provide the answers, and I don't think it can. I think in some ways it's dangerous for the lesbian and gay community to use biology as a proxy for that debate."

THEY KNOW ABOUT DARYL BEM at NARTH and refer regularly not only to his successful heterosexual life but to his contention that sexual orientation is a developmental, rather than a biochemical, matter; that it originates in the extent to which children are gender-conforming (70 percent of gay people report that as children, they felt more like members of the opposite sex than their own), a theory that NARTH claims as support for its therapy with young gender nonconformists. They are also familiar with Lisa Diamond's work; I first heard of it in Dean Byrd's seminar. And they know about people like the Wet Spotters.

"We know that straight people become gay," NARTH president Joseph Nicolosi told the group gathered in Orlando. "So it seems totally reasonable that some gay and lesbian people would become straight. The issue is whether therapy changes sexual orientation. People grow and change as a result of life experiences, especially personal relationships. Why then can't the experience of therapy and the relationship with the therapist also effect change?" (Diamond called this interpretation a "misuse" of her research. "The fluidity I've observed does not mean that reparative therapy works," she told me. And Bem said, "I know hair transplants work, and I'm not sure that the NARTH prescription will.")

They even know about me here. In 1997, I wrote a paper on the deletion of homosexuality from the *DSM,* arguing that the decision, while politically sensible, was hardly scientific. The paper is cited on NARTH's Web site, and when Dean Byrd asked me to lead off the introductions that started his seminar, just mentioning the title of the paper went a long way to countering the suspicions raised by the blue-state, Jewish origins plastered on my name tag.

But the charm of my credentials eventually only takes me so far. When I come out at NARTH (as a journalist), my confreres suddenly become less candid. Byrd whips out his cell phone when I approach him for a conversation that I'm hoping will lead to an interview. Kermit Rainman says he won't talk any further on the record without anonymity, which I grant him—only to discover that his story is plastered all over the Internet. A twice-married anesthesiologist suddenly clams up when I ask him to tell me more about confessing his gay past to his lesbian daughter. "I don't want to be some kind of poster boy," he says and bustles away before I can get his name. And the ex-Muslim, ex-gay, born-again doctor from Azerbaijan breaks our date to see *Borat* in order to have dinner with Byrd. Too bad; I'd have loved to get his impression of the wrestling scene.

But after Byrd closes the conference with a rousing speech urging his colleagues to "insist on a place at the table and be prepared to take it," I decide to do exactly that, tagging along with a group for a postconference lunch at Tony Roma's. We're waiting for our meals—a couple of students; the Azeri doctor; the anesthesiologist and his second wife; a Swiss German psychiatrist; a therapist from Colorado; and a couple of college kids, all of them ex-gay—and the Americans are telling the foreigners about some of the cultural life here.

"The singer I really want to see," one of the college kids says to Lukas, the Swiss doctor, "is Dolly Parton." Lukas has never heard of her.

"Dolly Parton," the other kid says. "There's two good reasons to see her." He looks around the table, but no one is taking the bait. Finally, he holds his hands way out in front of his chest, cupping large imaginary breasts. "Wouldn't mind seeing her at all," he says.

The waitress finishes delivering our meals. She's a professional flirt, and I wonder how she's taking our table's apparent oblivious-ness to her charms when the Coloradoan says, "She sure is cute." Silence. "Our waitress," he says again, "she sure is a cute one."

A small ripple of assent finally passes around the table.

All of this forced machismo reminds me of something else I've seen this weekend: young men standing close to one another, locked in intense conversation, sometimes even hugging (part of being ex-gay, the reparative therapists say, is learning to be same-sex affectionate without being same-sex sexual), but always work-ing hard, it seems to me, to keep some distance, to prevent desire from flowing in the wrong channels, if not to keep it in the right ones. When a couple of men, sparks flying, get in the elevator together one night, it's obvious that this isn't always possible, that "ex-gay," at least to judge from the ex-gays I've been surrounded by all weekend, is *not* totally heterosexual.

But that doesn't mean it isn't real. Ex-gays may simply be peo-ple living in bad faith, but they may also be an emerging sexual minority with something to prove; not people forced into the closet by an oppressive society, but people who—no less than the Wet Spotters, and with a no less political purpose—are breaking the boundaries imposed by the idea of sexual orientation as something inborn and permanent. When they talk about their martyrdom, the way that they are victims of political correctness, silenced by a science that gay activists hijacked in 1973 and have exploited ever since, it's surely a page right out of James Dobson's playbook, but NARTH is right on at least one count: the complexity of sexual orientation surpasses the certainties of biology. To the extent that proponents of gay rights rest their claims on a scientific founda-tion, NARTH's strategy is bound to pay off. Gay activists will then be left to build on other sources of public sympathy, none of which has the appeal of science. After all, if sexual identity is more like religion than race, a matter of affiliation rather than of birth, and fluid rather than fixed, then finding a different basis for popular support, as well as for legislative and judicial protection, means directly confronting something that Americans are perpetually confused about: the nature and the boundaries of pleasure.

NARTH is perfectly positioned to exploit this confusion by arguing that sexual orientation is subject to change, is fluid like a river whose direction can be influenced by environmental conditions, and then claiming that certain courses are less healthy than others. That's how the NARTH-ites justify working hard to make the world a less hospitable place to gay people. They oppose gay marriage and adoption, among other measures, not because they abhor homosexuality but because a gay-friendly world is one in which it is hard for gay people to recognize that they are suffering from a disease for which they should seek treatment.

Of course, in deploying medical language to serve its strategic interests, NARTH is only following the lead of Kertbeny and Hirschfeld, the original gay activists, and their modern counterparts who, despite minimizing the importance of biology, resort to scientific rhetoric when it suits their purposes. "People can't try to shut down a part of who they are," said Sean Cahill. "I don't think it's healthy for people to change how their body and mind and heart work."

We rely on doctors to tell us what is "healthy," but medicine will always seek to change the way people's bodies and minds and hearts work; yesterday's immutable state of nature is tomorrow's disease to be cured. Medical science can only take its cues from the society whose curiosities it satisfies and whose confusions it investigates. It can never do the heavy political lifting required to tell us whether one way of living our lives is better than another. This is exactly why Karoly Kertbeny originated the notion of a biology-based sexual orientation, and, to the extent that society is more wtolerant of homosexuality now than it was 150 years ago, that idea has been a success. Since Kertbeny established his sexual taxonomy, we have only gotten more dependent on science to resolve moral dilemmas, more in the habit of explaining our deepest and most intractable inclinations as necessities imposed upon us by biology. But the ex-gay movement may be the signal that this solution cannot suffice; that Kertbeny's invention has begun to outlive its usefulness; that sexuality, profoundly mysterious and irrational, will not be contained by our categories; and that it is time to find reasons other than medical science to insist that people ought to be able to love whom they love.

4

SCHIZOPHRENIA:
IN THE KINGDOM OF THE
UNABOMBER

THE FIRST TIME I GOT A LETTER from the Unabomber, I had my wife open it.

I was at work, the letter had come to my house, and neither of us wanted to wait to see what Ted Kaczynski, whose outgoing mail was by then inspected by the U.S. Bureau of Prisons, had to say. Sealed in a number 10 envelope, the letter was addressed in the careful block capitals that the post office says will guarantee maximum efficiency. He even put his return address, in the same print, in just the right spot. Kaczynski had considerable experience with the post office, so he knew that it works for you only if you work with it.

The first letter, which arrived in June 1998, had not come unbidden. Six months earlier, just after he'd pleaded guilty to the Unabomb crimes, I'd written Kaczynski a letter. Although I had paid close attention to his case for nearly three years, from his emergence as a composite sketch demanding space for his manuscript in a national publication to his arrest, incarceration, and abortive trial, my letter wasn't fan mail. Instead, it was a pitch. He was at the height of his fame, and I wanted to write a magazine profile, perhaps even an entire book, about him. I was, I told him, primarily interested in the way that he had been labeled a madman, a paranoid schizophrenic, when nothing that had appeared in the media about him (except the psychiatrists' conclusion that he was crazy) supported that view.

My prospective subject was interested enough in the project to ask, through his lawyer, for more information about me. So, during the spring, I wrote Kaczynski a short autobiography. I told him about my therapy practice and my teaching, even a little about my personal life, and I sent him some of my academic writings: two articles and a book. I heard nothing directly, and in mid-May 1998, after he'd been transferred to the Supermax prison in Florence, Colorado, I sent him a gentle reminder of my existence. His first letter came in response.

The letter was four pages long. It was written in a precise and blocky print on college-ruled paper. There were no signs of erasures or corrections. The prose didn't so much flow as march steadily from the beginning of an idea to its end, a flawless parade of logic. The letter was courteous, reasonable, and promising.

It would have been easy, in fact, to forget who my correspondent was, if it weren't for the question with which he started our exchange.

> Do I infer correctly that you believe that there is no such thing as objective truth? That all truth is relative to culture, values, attitudes and the like? If this is what you believe, then how would you answer the following objection? Consider the truths of nuclear physics. They tell us that if a device is constructed using a certain quantity of plutonium in such-and-such a configuration, a nuclear explosion will

be produced. Anyone within range of that explosion will be incinerated regardless of his culture, values, attitudes, etc. So, just what can you mean by saying that the truths of nuclear physics are relative?

Kaczynski wanted to know about my relativism. But I wanted to know why he chose that particular example.

Even more, I wanted to know how the person who had fashioned this note, with its politeness and sensitivity, its levelheaded clarity, its measured expression of frustration—how this person had spent seventeen years of his life perfecting a technique for building bombs and delivering them to people he didn't know. It's hard to square murder or other depraved acts with rationality and the other hallmarks of mental health. But this might be more about the way that medicine, and in particular psychiatry, goes proxy for morality than about the psychopathology of Ted Kaczynski or any other murderer.

Kaczynski had thought about this, which is probably why he responded to my pitch. He had a personal stake in this question, after all. He wanted to prove that he wasn't mentally ill. Or, to put it another way, that his behavior needed to be debated in the old-fashioned terms of good and evil.

PEOPLE TEND TO THINK that we psychotherapists practice a quasi-medical art, but having done it for many years, I can tell you that our job is much more like ministry. We don't preach the word of God—indeed, if we are decent at our profession, we don't preach all—but we do engage in a kind of moral argumentation. Because what is a therapist if not someone who helps people to improve their lives, and what is improvement if not a move toward a better life? And how do we know what is better? We can invoke the language of health and illness, but that doesn't really work unless you extend those terms beyond their traditional use. We therapists dispense our notions of the good (and the bad) from the deep cover of medicine; we tell you that you should listen to us, answer our questions, engage in the kind of introspection we encourage because we know, presumably through science, that it is *good for you*—and not simply because we believe, due to our faith or values, that it is *good*. It's an excellent deal in this respect: the power of the priesthood without the vows of poverty, and without having to clarify or justify our theology.

This idea—that all of human depravity can be reduced to mental illness—is not solely the invention of psychology, but it is no coincidence that it began to emerge toward the end of the nineteenth century, at just about the same time that Nietzsche declared that God had died. Not everyone believed this, of course, but Nietzsche was onto something important about the modern world: that as scientists pried more secrets from nature, revealing belief in the supernatural as a form of ignorance, they also replaced priests as the arbiters of ultimate truth. As the scientific gaze was turned inward, this had unforeseen consequences, the most important of which is perhaps the tendency, by now nearly a reflex, to attribute what worries or disgusts or inflames us about ourselves or other people to psychopathology and to turn disagreement into diagnosis.

Consider this excerpt from Ted Kaczynski's journal. It relates an incident from 1966, when, as a graduate student at the University of Michigan, he'd consulted a psychiatrist about an embarrassing personal problem: he may have been a brilliant, Harvard-trained mathematician and a decent-enough-looking lad, but Ted Kaczynski just couldn't get laid. He didn't consult the shrink to brush up on his interpersonal skills or plumb the depths of his ambivalence toward women, however. Rather, he had decided that the next best thing to getting a woman was *becoming* a woman. (This is not an uncommon motive in men seeking sex-change operations.) But while in the waiting room he changed his mind and instead of pleading his case as a transsexual improvised a tale about anxiety over the possibility of being drafted. As he left with humiliation piled on top of shame, the future Unabomber had an epiphany.

> As I walked away from the building afterwards, I felt disgusted about what my uncontrolled sexual cravings had almost led me to do and I felt humiliated, and I violently hated the psychiatrist. Just then there came a major turning point in my life. Like a Phoenix, I burst from the ashes of my despair to a glorious new hope. . . . I said to myself why not really kill the psychiatrist and anyone else whom I hate. What is important is not the words that ran through my mind but the way I felt about them. What was entirely new was the fact that I really felt I could kill someone.

Now, even if you don't know what eventually became of Ted Kaczynski, your first response to the sexual confusion, the violent impulses, the grandiosity, and the chilling conclusion contained in this journal entry is something like, "Wow, that's really sick!" Of course, we mean this colloquially, but this is the intuition that lies behind the often-lamented expansion of the insanity plea: that people who behave badly must be ill.

It's also what lies behind both the pathologizing of homosexuality and its subsequent deletion from the *DSM*—the subject of one of the articles I sent him in my introductory package. What changed over time was not the mental health of gay people, but instead the notion that there was something wrong with having sex with someone of your own gender. The "disease" had been invented to express the moral disapproval of sexual behavior in a less punitive fashion than, say, placing homosexuals in prison. Once that disapproval began to evaporate, particularly among the educated elite who have their hands on levers of social control such as the *DSM,* the disease had outlived its usefulness. Indeed, the "science" behind the deletion is notoriously weak. It consists largely of the results of psychological testing that explicitly *doesn't* test for homosexuality. It turns out that if you eliminate the homosexuality scales on tests of psychopathology, as sympathetic researchers began to do in the late 1950s, then the homosexuals who take the tests don't show up with pathology in numbers greater than the rest of us. If you're ever looking for evidence of the logical impossibility of proving a negation, you need search no further than this body of work, which proves only that homosexuality isn't associated with mental illnesses other than homosexuality. That's what the psychiatrists who objected to the deletion of homosexuality recognized: that by this logic anything could happen—even schizophrenics advocating for the deletion of schizophrenia.

This, of course, doesn't make deletion a bad thing but rather a good thing dishonestly attained. It was, as the paper I sent to Kaczynski stated in the title, the right answer for the wrong reason. I don't know whether he thought that the scientists who declared that homosexuality was not a disease were telling a noble lie or just a plain old lie, but when he responded to my article and the letter that accompanied it with a twenty-page outpouring, I wasn't surprised. He wasn't a total fan of my work.

> As a freshman at Harvard, I took Humanities 5, History of Philosophy. As a result, I concluded that philosophy was a lot of bullshit, and I've never found any reason to change that opinion. . . . In most philosophical works what is of value is so mixed up with bullshit that it takes more trouble than it's worth to dig out the genuine insights. Your article provides a good illustration.

But Kaczynski did allow that it was "possible to carry on a rational discussion" with me, and that he'd be pleased to continue to do so. The use of psychiatry to make a political point was a subject even closer to his heart than I had guessed.

NOT ALL OF HIS SECOND LETTER, which arrived in early July, was about the nature of mental illness, nor was it as formal as the first. In fact, it was addressed, "Dear Gary," and signed, "Best regards, Ted Kaczynski." (From then on, we were on a first-name basis.) It was personal in places, revealing even, as when he wrote:

> I used to have bad dreams in which I would see myself and my cabin isolated on a tiny little patch of land surrounded by a gigantic shopping center.

It was expansive and included five footnotes, which ranged from simple amplifications of what he was saying to quibbles with me about my interpretation of early Christian martyrdom. But he kept coming back to the subject of psychiatric diagnosis.

You couldn't blame him for this. After all, if psychiatry didn't cloak its moral judgment in specious diagnostics, he might not be in his current position: left to rot in Supermax, where his bed and table are made out of molded concrete and exercise takes place in a kennel.

Instead, he'd be dead or at least under a death sentence.

To understand why my paper got a twenty-page rise out of Kaczynski, you have to know a little Unabomber history.

Kaczynski's lawyers knew a hopeless case when they saw one. There was a warehouse of evidence against him: bomb-related hardware, journal entries lamenting his failures and applauding his triumphs, various eyewitnesses to his where-abouts. Worse, the federal government had a new death penalty, and the Unabomber seemed a fitting first target: he'd committed heinous crimes, embarrassed the FBI by eluding it for almost two decades, and seemed entirely unrepentant. To his lawyers, this meant that there was only one possible plan: to find a defense that would minimize their client's chances of getting executed. But to Kaczynski, this was an end that served the lawyers more than their client. Furthermore, as he wrote to attorneys whose support he sought after he had been convinced, it just wasn't fair:

> The principle that risk of the death penalty is to be mini-mized by any means possible . . . is very convenient for attorneys because it relieves them of the obligation to make difficult decisions about values or to think seriously about the situation and the character of the particular client.

The problem, in Kaczynski's view, was that the single course that would save his life was to turn to the psychiatrists and make himself out to be a crazed killer. After all, if you're going to kill in cold blood, which is what a juror is asked to consider, your victim had better be a villain and not someone to whom you can, as we therapists say, relate. Fortunately for defendants with good lawyers, there is no end to my profession's ability to commonly denominate the most heinous act or the most loathsome personality: Charles Manson had a mother, too. Thus is revulsion turned to empathy, and all of transgression's horror is reduced to the banal recitation of trauma that everyone might share.

So the defense rounded up its investigators and psychiatrists to prove that this hermit, with his poor hygiene and inscrutable mailing list, was a nut. They even arranged, in a strange fulfillment of Kaczynski's bad dream, to bring his cabin to Sacramento for the jury to examine.

"You've got to see this cabin to understand the way this man lived," said Quinn Denvir, his lead defense lawyer. What you would see, Denvir explained to the press, is the external manifestation of a demented mind. "The cabin," he said, "symbolizes what had happened to this PhD Berkeley professor and how he came to live. When the people think about his case, they think about the cabin."

Back in the early 1980s—and here's another way that Kaczynski and I were made for each other—I retreated to my own little cabin in the woods, a place I built on an old family farm and where I lived without benefit of modern conveniences. Most people thought I was at least a little nuts to live this way. If I had had legal trouble, which would have more likely been about growing dope than about building bombs, I don't think I would have wanted my lawyer to be among these doubters. But that was Kaczynski's situation. His lawyers wanted to save him from execution, and to do so they were willing to turn the better part of his adult life into a case study. Kaczynski didn't want his life saved that badly.

He did manage to have some fun with the psychiatrists and the psychologists they sent his way under various covers: to help him with his sleeplessness in the noisy jail, a condition that one doctor called Kaczynski's "over sensitivity [sic] to sound"; to give him tests that might prove that he was neurologically intact; and to assist in the preparation of his defense. They all came back empty-handed, with no raving lunacy or other florid symptom to report. Kaczynski refused to talk about his feelings, terminated interviews when clinicians started to talk about his mental illness, and told his lawyers repeatedly that he would not cooperate with their defense.

Kaczynski had opted out of American culture in the late 1960s, at just the time that everyone was learning to speak the language of therapy, but it wasn't ignorance that kept him from a crying confession of psychic pain. He knew just what the shrinks were up to, not only in terms of his trial but also in the larger sense: they were trying to tell his story in their language, which was unacceptable to him.

Many clients refuse to accept the therapist's authority, but most are reduced to the squirming prevarication we call "resistance": missing appointments, changing the subject, disavowals of feeling. Kaczynski, however, just up and said it. Dr. David Foster, who met with him five times in 1997, wrote, "Early on in our sessions, he looked me in the face and said, 'You are the enemy.'" Kaczynski wrote to me about this comment:

> I was simply laying on the table in a civil, or even friendly way, as a matter that needed to be taken into account in our discussions, the fact that Foster and I were on opposite sides of the ideological fence, that he as a psychiatrist was an important part of the system I abhorred, and that he was in that sense an enemy.

In Foster's version, Kaczynski's candor reflected "his paranoia about psychiatrists." This itself was part of his "symptom-based

failure to cooperate fully with psychiatric evaluation." There
are no principles in this world, only symptoms; no politics, only
pathology. Of course, Foster, like all the others, knew what everyone else knew: that this man was the Unabomber, so he must be
crazy. The fix was in from the beginning. Even his defense lawyers were in on the game, ultimately arguing that Kaczynski's disagreement with them about the mental-defect defense was more
evidence of his mental defect. No wonder they all thought he was
paranoid—they were out to get him.

WHEN IT BECAME APPARENT that his lawyers were going to go
ahead with the mental-defect defense, Kaczynski stopped the
courtroom proceedings and asked to represent himself. But
the judge, Garland Burrell, insisted that he had to determine that
Kaczynski was competent to do so, so in order to avoid a psychiatric defense, Kaczynski had to submit to a psychiatric examination.
It was conducted by Sally Johnson, the same psychiatrist who had
found John Hinckley, Ronald Reagan's would-be assassin, insane.
She interviewed Kaczynski for twenty-two hours and determined
that he was a paranoid schizophrenic.

   This in itself was nothing new; it had been the conclusion of
all the other doctors, but they had to coax the diagnosis either out
of Kaczynski's known history or his current orneriness. They had,
for instance, taken the fact that he used his own composted shit to
fertilize his garden (a practice not quite as unusual as it sounds;
there's even a name for it: *humanure*) as evidence that he suffered
from "coprophilia," an unhealthy interest in feces. His hardscrabble, Third World life showed a lack of self-care. And his failure
to accept that he was cruelly deranged was "anosognosia," the
condition of being too sick to agree with the psychiatrist, which
is a hallmark feature of schizophrenia and a word to bear in
mind the next time you disagree with a psychiatrist. But Johnson
needed to do no diagnostic conjuring. In twenty-two hours, she
had taken the measure of the man, gotten a full frontal view of

the Unabomber, and she'd concluded that he was really and truly crazy, at least provisionally.

Of course, according to the current *DSM,* to earn that diagnosis you have to do more than think psychiatrists are the enemy. You also have to have delusions, and Johnson thought she had found them, as she wrote toward the end of her report:

> In Mr. Kaczynski's case, the symptom presentation involves preoccupation with two principle [sic] delusional beliefs. A delusion is defined as a false belief based on incorrect inference about external reality that is firmly sustained despite what all most [sic] everyone else believes, and despite what constitutes incontrovertible evidence to the contrary. [I]t appears that in the middle of late 1960s he experienced the onset of delusional thinking involving being controlled by modern technology. He subsequently developed another strong belief that his dysfunction in life, particularly his inability to establish a relationship with a female, was directly the result of extreme psychological verbal abuse by his parent [sic]. These diseases were embraced and embellished, and day-to-day behaviors and observations became incorporated into these ideas which served to further strengthen Mr. Kaczynski's investment in these beliefs.

So here was the final proof that Kaczynski was crazy: he thought technology controlled his life, and he believed that his parents had made mistakes that had made his life miserable.

As delusions go, these are problematic. Technology surely mediates our lives, even if it does not control them outright. And the question of parental abuse is an epistemological black hole. Rarely, if ever, does a therapist get corroboration (or incontrovertible contradiction) of a client's claims that he or she was subjected to bad parenting. Indeed, it is often the case that therapists "help" their skeptical clients to see that they were abused.

Dr. Johnson would have had a partial answer to these objections: it wasn't what Kaczynski believed so much as the tenacity of his belief that was troublesome. Try as she might, she couldn't persuade him of the folly of either of his "delusions." "When challenged on the initial premise [of either belief]," she wrote, "he appeared perplexed and it was evident that he did not challenge the belief system on his own regardless of existing evidences." Even worse, "He does not challenge [his beliefs] in response to new information."

Johnson never made good on her promise to give Kaczynski her notes from their interviews. That's too bad, because it would be interesting to see just how this conversation between two people who disagreed on basic premises went. One thing is clear, though: there was no way for Kaczynski to respond (other than agreeing with Dr. Johnson that technology wasn't such a bad thing and that his family was functional) that would not reinforce his diagnosis. What the psychiatrist overlooked, however, was that by her logic—in which their disagreement was about not politics but about reality itself—one of them had to be crazy. But it might not have been Kaczynski.

So Kaczynski was found guilty of schizophrenia but still competent to stand trial, which meant that he was competent to defend himself. Then Judge Burrell, whose knickers had been twisted by this mathematician's unassailable logic and dogged insistence on obtaining the protections of the system he hated, played his last card. When he denied Kaczynski's motion to represent himself, Burrell made no use of Johnson's report; he simply ruled that the motion had come too late, even though Kaczynski had repeatedly indicated that he was ready to proceed immediately. He had to go through with his lawyers' defense.

The Unabomber had been bamboozled. Now he had the worst of both worlds: the psychiatric exam he had never wanted, and the certain prospect of hearing its findings reiterated in open court. He felt that he had no choice but to plead guilty. And five months after he made this choice, when Kaczynski got my paper on the bankruptcy of psychiatric diagnosis, he must have thought that even if I didn't already know all that had happened to him, I would probably understand and believe him when he said he'd been bushwhacked. That might be why he wrote me a twenty-page letter in response. There was someone inside the industry who wouldn't think he was crazy simply because he didn't like psychiatry. He must have figured he could use such a person, and he turned out to be right.

MORE WAS AT WORK HERE than grudging respect and my own ambition. I thought Kaczynski had something important to say, something worthwhile, something that could stand to be put into both historical and psychological context—notably, that technology was not simply an assemblage of tools, awaiting our use, wise or foolish. It was a way of being in the world, and one that had some obvious problems. In particular, it seemed to leave us fully aware of, but unable to do anything about, the way our devices alienated us from one another and the natural world and, as a result, threatened great peril. Technical progress had trumped all other ends to which humanity might be put, had made us slaves, in other words. In *Industrial Society and Its Future,* aka the *Unabomber Manifesto,* Kaczynski argued that technology didn't take away our freedom forcefully. Rather, enchanted by its near-magic powers, we had become collaborators in our own enslavement.

None of this was original to Kaczynski, although it had probably never appeared in the *Washington Post* before. The Industrial Revolution has always had its naysayers, artists and philosophers and social theorists who question what it is doing to us. Crucial among these questions, at least for a psychologist, is how we do anything about it. William Blake, an early antimodernist, captured this process with his image of "mind-forg'd manacles," shackles that are so compelling and comfortable that they become undetectable and show up only to people the way they are, without knowing it, imprisoned by their own unacknowledged history. But some cases of self-imprisonment are harder to understand

and point out than others. And the one that Kaczynski noted is perhaps the hardest of all. Technology not only helps us to accomplish things, with the occasional failure or accident or frustration; it also constructs us as the kind of people who are hard-pressed to be sufficiently critical of technology.

Perhaps the mental health industry, as Kaczynski implied, is inescapably another of the sorcerer's apprentices. That's one way to explain the difficulty of understanding, at least in psychological terms, this central mystery of technology, the way it seems to keep us blind to itself. But the fact is that no one really understands how we can listen to another report about global warming even as we drive our cars, festooned with "Save the Earth" bumper stickers, to fetch a loaf of bread. No one really knows how we sustain this level of what psychologists call cognitive dissonance or why we barely perceive it. Neither can anyone explain why we are not wracked by guilt and anxiety or at least repelled by our own bad faith. And because we (psychologists, that is) don't really understand these things, we can't do anything about them, even if we want to. Such has always been the problem with thoroughgoing indictments of modernity: they're long on critique and short on solution.

The *Manifesto*'s proposed therapy parted ways with this aspect of antimodernism:

> The only way out is to dispense with the industrial-technological system altogether. This implies revolution, not necessarily an armed uprising, but certainly a radical and fundamental change in the nature of society.

And it offered a very loose treatment plan:

> It would be better to dump the whole stinking system and take the consequences.

My philosophical kinship with Kaczynski—in which I don't think I was by any means alone; as Robert Wright wrote in *Time,* "There's a little bit of the Unabomber in most of us"—stopped short of this let-the-chips-fall confidence. I like the fact that I don't have to worry about getting smallpox, and I'm not quite willing to say that the whole system ought to be jettisoned or the citizenry rallied to arms by random violence, as Kaczynski evidently wanted.

But the fact that he was a killer perhaps only increased my interest. Was it possible that Kaczynski's moral depravity was understandable as the snapping of a weak link in a chain pulled too tight? Was it possible that his terrorism was only the leading edge of a series of even more desperate acts to come, as that cognitive dissonance came to be less and less tolerable? That his very character seemed to bear the imprint of large social and historical forces, that he seemed to know what those forces were, and that he was very, very famous—all this made him an irresistible subject.

Not that any of these qualities (except, of course, his fame) were part of Kaczynski's public image, at least not anymore. He had once been a potent political figure for a blink of an eye. But celebrity culture doesn't just hand out names for free. Kaczynski, having gotten famous by unsanctioned means, had to pay the price. He couldn't be forgotten, and he certainly couldn't be bought out of his beliefs. So he had to be turned into kitsch. And, to make things worse, his fashioning as a pop-culture trinket was largely brought about by his own lawyers, at least according to William Glaberson of the *New York Times*:

> The shift in public image, which began with Mr. Kaczynski's arrest for carrying out an 18-year campaign of bombings that killed 3 and injured 28, accelerated after his lawyers said he was a delusional paranoid schizophrenic who believes people have electrodes implanted in their brains.

To keep Kaczynski safe for democracy, his license to seriousness had to be revoked. If he's crazy, after all, then he can be famous without being meaningful, his unsettling denunciation of modern technology reduced to the entertainment of a lunatic's raving.

And who, besides the lawyers, was responsible for this outcome, this down-the-rabbit-hole reversal of logic whereby a rational, if contentious, belief—that there's something wrong with the way technology has colonized our landscapes, both interior and exterior—becomes the mark of insanity? Therapists, of course, the people who are trusted, for no particularly good reason, with the authority to decide who is a genuine apostate and who is just plain nuts, whom we should listen to and whom we can dismiss. The first person who might have predicted this outcome was Kaczynski himself, who worried a lot more about therapists' inability to distinguish pathology from dissent than about their implanting electrodes in his brain. The culture indulged his anxiety, and its agents were my own colleagues.

THAT JULY AND EARLY AUGUST brought more letters. They were dense, carefully argued, and full of promise, even at one point indicating that I could come to visit him. They covered ground from Russian history to Hobbes's *Leviathan* to Desmond Morris to the hygiene habits of Indianans, who, he said (after apologizing for not remembering the source of this information) once had a law prohibiting bathing in the winter (which he thought indicated the Hoosiers' wise conclusion that pneumonia was a more important problem than body odor).

The letters also took a surprising and unsettling turn toward the personal. He offered information about himself and asked detailed questions about me.

I found myself intrigued in a way I hadn't expected and unable to resist making some clinical observations about him. He was, I thought, just as complicated and full of self-contradiction as the rest of us. While he tried to live a life of complete consistency between his beliefs and his actions, in some ways he embodied the biggest opposition of all. He was at once a mathematician, a man of science, entirely convinced of the superiority of reason as a means of negotiating the world, and a savage critic of rationality's greatest achievement: technology. It's impossible to divorce Descartes' ego cogitating its way to certainty from Henry Ford's Model T slipping down the conveyor belt—both grow from the desire to hold the world firmly in our grasp, to make it yield to us. Most of us see the resulting nest-fouling problem. Kaczynski saw it, too, but he seemed unable to turn this infinite loop of alienation into the wry irony the rest of us are so good at. It just pissed him off.

A differently constituted man might find the tension of being stretched across this great rift of modernity unbearable. Perhaps this is why the psychiatrists who evaluated him found him to be schizophrenic even though, at least in their presence, he never behaved like a schizophrenic. Maybe they divined the desperate, irreconcilable conflict in his politics and concluded that a man unable to gloss over this problem like the rest of us *ought* to be crazy.

Kaczynski had also been thinking about how to make use of our contact. He wanted me to read the book he'd just finished writing, comment on it to him, and then consider the possibility of interviewing him and his family, to come up with a fairer assessment of the Kaczynskis.

Truth Versus Lies arrived in early September, a 548-page typescript. It was Kaczynski's point-by-point, fully documented refutation of all the unflattering things the media had said about him. His thesis was that his brother, David, and his mother, Wanda, rather than acknowledging the Kaczynski family dysfunction, had portrayed Ted to the national press as mentally ill. A willing and gullible media had then amplified this account until Theodore Kaczynski had become, in the public eye, just another lunatic.

Kaczynski is unrepentant in the book, addressing the Unabomber crimes only obliquely and often providing details that can only make people like him less—as, for instance, his dispute of a news account about a dirty limerick regarding a coworker that he'd scrawled on the workplace wall. Kaczynski's rebuttal was that he'd scrawled the limerick on a vending machine.

I suppose the all-trees-and-no-forest approach of *Truth Versus Lies* could be read as evidence that Kaczynski is mentally ill. But surely it is not evidence of neurochemical explosions of schizophrenia and the resulting disorganization of mind. Quite the contrary. The book is remarkable for its controlled tone, the steady focus it brings to bear on a sprawling archive of personal and public history. And underlying it is a method that is both coherent and quaint. It conjures a world, part nostalgia, part desperate hope, in which great issues are discussed in measured tones and brought to incontrovertible resolution by reasonable men in dark-paneled rooms. In that imagined world, people will look at the facts and soberly reconsider their conclusion that a man who lives in the woods and sends bombs through the mail to people he doesn't know, who renounces the bounty of industrial civilization and fertilizes his garden with his own shit, must be crazy.

So it's not entirely true that *Truth Versus Lies* is devoid of self-flattery. It's just that one of the qualities most worth having, in Kaczynski's view, is rationality, which he has by the bucket. I think the only hero he ever wanted to be was Rudyard Kipling's hero in his poem "If," the one who kept his head while all about him were doing otherwise.

He wasn't even blaming others for what he had done. Even if he claimed that he was just another abused child, he offered no excuses. He seemed content to tell the truth, to withstand its reflection upon him, and to settle for the cerebral satisfaction of possessing the facts. After reading the manuscript, I told Kaczynski that his approach to the problem wasn't likely to change any minds and that his chosen method was like using Euclidean geometry to argue with a hurricane. I added that I thought the manuscript would backfire and give new currency to the image he was trying to discredit. And without saying so, I declined the opportunity to serve as the Unabomber family shrink.

BUT I WAS STILL INTERESTED in writing about him, an interest that had by then morphed into a contract with a national magazine for an interview, the first ever with Kaczynski. There was, of course, one slight problem. He hadn't yet consented to give me face time; in fact, I hadn't really asked. But *Truth Versus Lies* provided my opening. At the end of my critique, I told Kaczynski that if he really wanted to redeem his public image, he might consider allowing me to interview him for a national magazine, and that I happened to know of a magazine that was interested. I spent the next couple of months in hot pursuit of an audience with the Unabomber. An alternating current of come-hithers and get-losts ran from his cell to my mailbox, sometimes twice a week. The man of perfect logic was clearly confused about what to do, and as a result, he knitted and unraveled promises like Penelope.

Kaczynski's dithering was unfailingly polite, and he often apologized for it. It wasn't anything about me personally; in fact, he said, he liked me and thought I wrote well. But there was a perhaps insurmountable problem. He was afraid that even if I got all the facts right, I would still get him wrong. He understood that I wanted to be more than the Unabomber's amanuensis, that I had ideas of my own. And he was concerned that those ideas might make me misrepresent him. Of course, he allowed, it was possible that I might see things about him that he himself could not, that the error might be his and not mine. But it was also possible that I would be mistaken, would see things that weren't there, and would turn him into someone he was not. And this, he told me repeatedly, would horrify him.

For Kaczynski, it was impossible that more than one story could be true. Like any good empiricist, he was sure that the world and the people in it could be divided into the really there and the not. My credentials only deepened his worry that I would get his story wrong and then accuse him of bad faith for his disagreement. But the real issue was deeper: it was the inevitability that my values would seep into my account of him. His personality would then be no more than a platform for my own ideas, and he would be stuck with yet another story about him that wasn't true, only this time told with his own consent. After all, he hadn't gone into that cabin just to avoid an electric bill. He'd gone there to keep himself intact, away from the institutions—corporations, universities, psychology—that would make him into their own versions of him. He thought that he had the best command of the facts of his life.

But the prospect of having his name cleared continued to tantalize him, and he finally offered me an audition. "Publish one or more articles on the basis of the information you already have," he wrote. "When I've read them I'll reconsider."

As it happened, I had something ready for him—an article, intended to be the foreword to a book about Kaczynski's defense by Michael Mello, a former public defender who saw the debacle as yet another instance of all that is wrong with capital punishment. The article was mostly about Kaczynski's diagnoses. I argued that Kaczynski's insistence on living his low-tech life, his hatred of the incursions of the modern world into the Montana woods, not to mention his aversion to psychiatrists, could only be symptoms if one already assumed he was delusional. The psychiatrists had assumed their conclusions, and without this fallacious reasoning, his attitudes and actions, which were undoubtedly deviant, were no more inherently pathological than, say, the claims of certain women that they are married to God and that they must wear strange clothing and live in convents to uphold their marriage vows. None of which was to say, I hastened to add, that Ted Kaczynski was not mentally ill; one can't, after all, prove a negation. But it was clear that the diagnosis was ill-founded and thus deeply suspect.

Kaczynski responded almost immediately. "Dear Gary," he wrote,

> Yesterday evening I read your Foreword. I think you did a superb job on it. I have objections to some details, but I don't think we'll have any difficulty in working those out, and the Foreword as a whole is excellent.
>
> On the basis of this Foreword, I would be willing to have an interview with you as soon as you like.

Of course, there was still a caveat. Kaczynski had recently been in touch with a lawyer who was considering taking his case. In lawyerly fashion, this man had advised him to curtail all contact with the outside world. So the final decision would have to await a discussion between the two of them.

What Kaczynski didn't know was that I already knew this lawyer, Richard Bonnie. Bonnie had read the article I'd sent for my audition, and told me that it had figured into his willingness to consider taking on the case. It had helped to convince him that at least insofar as the psychiatric evaluations were concerned, it was indeed possible that Kaczynski had been unfairly pressured into his plea bargain. Bonnie, through Mello, knew of my potential article for a glossy magazine and my negotiations with Kaczynski about an interview. He saw the value of such an interview, under properly controlled circumstances, to an appeal based in part on the injustice of Kaczynski's diagnosis. So, he told me, he would tell Kaczynski that he ought to go ahead with it, and that we would be going out to see him in January.

THE INTERVIEW NEVER HAPPENED. Kaczynski was just too skittish, too unwilling to put his public image in my hands, and when some other ambitious people with whom he'd been in touch denounced me to him, he seized the excuse to withdraw his invitation. I didn't mount a full-fledged defense of myself, even though the charges against me were trumped up. It was one thing to suck up to a serial killer in pursuit of a story and quite another to defend my own integrity to him and let him sit in judgment of me.

We exchanged a few letters over the winter, but the feel of a slack line in my hand was dispiriting, and, after a few notes, I let my end drop. I was surprised to find myself more relieved than disappointed—not only because I was no longer playing cat-and-mouse with the Unabomber but because, as much as I was sure that he was evil and not sick, I wasn't certain that I wanted my professional opinion about this to be his key to the death chamber.

But I'd forgotten something important. I forgot who I was dealing with. In April, I heard from Richard Bonnie. He'd decided not to take the Unabomber case, so Kaczynski had filed his appeal on his own. It was, Bonnie told me, 124 pages long and handwritten. And it included the article I'd sent him as an appendix, in support of his claim that he wasn't crazy and thus should never have been forced to choose between a mental defect defense and guilty pleas. You could look it up. It's Exhibit 9 of Theodore John Kaczynski's Pro Se Motion under 28 U.S.C. §2255 to Vacate Guilty Pleas and Sentences and Set Aside Convictions. This is how Kaczynski introduces me to the world:

> In respect to the ideological bias of the experts' reports on Kaczynski, see the essay by psychologist Dr. Gary Greenberg, attached as Exhibit 9.
> Kaczynski . . . emphasizes that Dr. Greenberg's essay contains certain errors of fact and erroneous conclusions. In attaching this essay to his petition, Kaczynski does not mean to express agreement with everything that the essay states or implies.

I guess he was still mad at me.

SO THE UNABOMBER GOT what he wanted: a report from a bona fide mental health professional attesting to his sanity. When I complained to him that he had used my paper without my permission, he was contrite, courteous, and understanding. He had made a mistake, he said; in the rush to prepare his defense, he'd let this stitch slip, and he sent me a heartfelt letter of apology that he said I could show to anyone who might accuse me of defending the Unabomber or aiding him to commit suicide.

In this, as in nearly everything he did and said to me, Ted Kaczynski was sane and sober and rational. Which is perhaps the most disturbing part of the Unabomber story: that a person who is sane, sober, and rational may do terrible things.

This statement is vexing only if we have already decided that behaving immorally is a criterion of mental illness. I believe this decision has already been made. It's implicit in the psychiatric case against Kaczynski: such specious reasoning can only bear scrutiny if it's what we already expect to hear. But a case like the Unabomber's forces us to look at his decision, and particularly at the way it puts my profession in charge of public morality.

Take my word for it, this is not a good idea. Not because my colleagues and I are scoundrels, although some of us may be, but because the mental health industry will reduce the political to the personal every time. It is our business to do so. Then we are stuck talking about health and illness instead of about right and wrong. Right and wrong, with their reach toward central questions of what it is to be human, are words worth discussing when it comes to serial killers, not to mention other important concerns, like what technology is doing to us and our world. Health and illness, aspiring only to scientific certainty, are, in comparison, hopelessly impoverished.

A society unaccustomed to understanding individuals' behavior as anything other than the result of their psychological states—their childhood traumas and neurochemical imbalances, say—cannot account for the political dimensions of everyday life. It cannot, for instance, raise the question of exactly what is wrong with what Kaczynski did. We perhaps could stand to be reminded of the public agreements that stipulate why we aren't supposed to kill, no matter the cause, and then perhaps we could decide what other people and practices are falling short of the standard that he violated. But the Unabomber case can't force this much-needed conversation if Kaczynski is merely a madman. Then it's enough to know that he is not one of us.

But he is. Indeed, *Time*'s assertion that "there's a little bit of the Unabomber in most of us" may not be all hyperbole. And it's not just the resentment inspired by the maddening little daily encounters—the questions that go unanswered because the computer is down or the thought interrupted by the cell phone or the privacy lost to the demographically precise database—that links us with Kaczynski. It's the knowledge of what lies behind these petty outrages. That's why, when we tell these stories to our friends, we cast ourselves as the heroes battling a wickedly impersonal world, struggling on the side of humanity against the machines and their feckless operators.

Because we know that something is not quite right out there. And it may be too much to assert, as the Unabomber did, that we are the trusties of modernity's prisons; it is certainly too much to kill random people for being collaborators. But it is not too much to say that the problems posed by technology are vast and complex and crucial, far outpacing the engineer's ability to repair a glitch or rethink a poor design. For it's not just the dangers and difficulties—the greenhouse effect and the nuclear waste and the extinction of various species—that ought to give us pause. Technology is etched deeply on our characters, perhaps as deep as our souls. In many ways, it gives us who we are: the kind of people who can flick a switch, hear the furnace rumble faintly in the basement, and take reassurance from its promised warmth without a moment's hesitation over where the oil came from or how it got here or what will become of its smoke; the kind of people who know the answers to all these questions, but what are you going to do, freeze? Move to a cabin in the woods?

We must wink at ourselves to get by. The little bit of the Unabomber in all of us may be an animosity toward an identity that is so thoroughly in the debt of bad faith.

The manufacture of the Unabomber as a crazed killer is highly efficient. It applies the balm of explanation to terrible events. It maintains a comfortable distance between him and us. It erases the nagging but crucial public questions raised by the story of a man unable to withstand the dissonance with which all of us must live.

5

BRAIN DEATH:
AS GOOD AS DEAD

THE LIVING ROOM of a small brick bungalow in central
Pennsylvania is as good a place as any to learn that death
isn't as simple as you thought. It's warm in here on an
icy night, and Rick and Kim Breach have just brought me tea.
They're both forty-three; he works at United Parcel Service,
"throwing boxes," and she works the phones for a big retailer's
customer service center. They're wearing matching sweaters, and
they sit close together on the couch, never out of physical contact. Sarah the dog scratches at the door of the room where she
has been sequestered for my visit. From the couch, they give me a
tour of the stuff in the room: the tatted lace curtains that are family heirlooms, the clock that never worked right, Rick's bronzed
baby shoes, and the VCR that, as Kim teases Rick, only she understands how to use. The Breach home feels like the kind of place
that a teenager would be glad to come home to after a trying day
at school.

Which is what their son Nicholas did until he got too sick to go to school. Nick is fourteen, and he has had astrocytoma, tumors of the nerve tissue in his brain, since he was six. He has been treated with surgeries and drugs, but a couple of months ago, an untreatable tumor turned up on his brain stem. The brain stem controls basic life functions such as breathing and temperature; it operates in the deep background, independent of the cerebral cortex and other structures that give us consciousness. The tumor leaves these higher functions alone as it incapacitates his body, so Nick is an alert witness to his losses, keeping an involuntary vigil over his own death.

Nick is lying in a hospital bed in the middle of the living room, between his parents and me. He can't move his eyes or head anymore, so when we speak, I have to move into his line of vision. The tumor has made it hard for him to move his mouth, but he's really good-natured about trying to talk to me. He even ventures a smile (he's wearing braces) while he tells me a story about his older brother, Nathan, and his dog. He apologizes for being so tired.

Before he dozes off, Nick tells me about what he hopes will happen to his body after he dies, how important it is that he be able to donate his heart and lungs and liver and kidneys and pancreas so that other people might live. Nick made this decision just after his medical team told him about his tumor. Bernadette Foley, Nick's social worker at Children's Hospital of Philadelphia, told me that the decision reflected a "maturity and sensitivity" and a wish to help others—something that she said Nick had shown throughout his eight-year battle with recurrent tumors.

"I've never been to a meeting like this one," Foley said. "The peace that came over the family and Nick was remarkable, and once it was out that this was the end and the decision was made about organ donation, Nick said he was happy. They all seemed to be happy." The decision was redemptive, she said. "In a way, it gave some meaning to his life."

Rick's feet are propped up on the bed, and Kim is adjusting Nick's pillows, scanning his face, she explains, for signs of discomfort. They're telling me about what they've had to do to uphold Nick's wish—and all the unforeseen complications. They had found out that organ donation is a high-tech affair. In most cases, the donor is someone with brain damage so severe that he requires a respirator to breathe, even though his heart continues to work on its own. A neurologist determines that the patient's brain has been irreversibly and totally destroyed and on this basis pronounces him dead. This condition is known as brain death. If the patient's family has consented to donation, he is left on the respirator, which, along with his still-beating heart, keeps his organs viable for transplant until they can be harvested.

Nick had decided that he wanted to die at home, with only palliative care, but the Breaches accepted that his wish to be a donor meant that Nick would have to be hospitalized at the very end. Their insurance company, on the other hand, balked at the change in plans—and the added expense—reminding them that they had already elected basic hospice care. Only after the family's state legislator and the regional organ procurement organization (OPO) got involved did the insurance company agree to pay. A plan was devised to keep Nick at home until the last possible moment and then to transport him to a hospital, where an informal protocol had been set up to help him become an organ donor.

But even with the logistical and financial arrangements in place, it was very unlikely that Nick would ever meet the criteria for brain death, that instead he would die from organ failure resulting from his brain stem's inability to regulate his core physiological processes. The fact that in order to become a donor, their son would have to die the right kind of death not only frustrated the Breaches; it also bewildered them.

"I'm so confused about this part of it," his mother said. "I don't understand why, if his heart stops beating, they can't put him back on a respirator."

Rick, for his part, understood that timing was everything but thought that the doctors would remain in control of when, as he put it, "the plug will be pulled." In reality, there is no such moment; to keep the organs viable, the respirator is left operating, and the heart keeps beating, until the surgeon removes the organs.

Neither of them has this quite right, but then again, brain death is really confusing, and not just to parents seeking some consolation for a horrible loss. It's confusing to transplant professionals, surgeons, neurologists, and bioethicists—the very people who work with it on a daily basis. Brain death is confusing because it's an artificial distinction constructed, more than thirty years ago, on a conceptual foundation that is unsound. In recent years, however, some physicians have begun to suggest that brain-dead patients aren't really dead at all—that the concept is just the medical profession's way of dodging ethical questions about a practice that saves more than fifteen thousand lives a year.

DENISE DARVALL WAS TWENTY-FIVE years old when she and her mother, Myrtle, were run down by a motorist on a Johannesburg, South Africa, street in December 1967. Myrtle died immediately, and Denise was taken to Groote Schuur Hospital with multiple fractures of the skull. She was placed on a respirator, but by five-thirty that evening, doctors had determined that the trauma and the subsequent bleeding and swelling had entirely destroyed her brain. Another patient at Groote Schuur was Louis Washkansky, a fifty-four-year-old man dying of heart failure. Christiaan Barnard, his doctor, had identified Washkansky as a good candidate for a heart transplant, a procedure that no surgeon in the world had yet tried to perform. At least one earlier attempt to secure a heart for Washkansky had been scuttled when the prospective donor's heart failed before his family gave consent. But Edward Darvall, having accepted the futility of further treatment for his daughter, agreed to let Barnard take Denise's heart for Washkansky. In the middle of that night, while Washkansky lay prepped for surgery in an adjacent room, Denise Darvall's respirator was shut off. It took twelve minutes for her heart to stop beating. Five minutes later, Barnard opened her up, and within less than a half hour, Darvall's heart was beating in Washkansky's chest.

Barnard didn't believe that he really had to shut off the respirator and wait those seventeen minutes, at least not for medical purposes. As far as he was concerned, Darvall was dead as soon as it was determined that her brain had been destroyed. But, as he later explained, he worried that the public would not accept this view and instead would believe that he had killed her by removing her heart. With the future of transplant surgery at stake, he said, "I did not want to touch this girl until she was conventionally dead—a corpse. . . . I felt we could not put a knife into her until she was truly a cadaver."

Barnard's concern, and the tension between pioneering doctors and a sometimes suspicious public, has become enshrined in transplant practice as a simple, unwritten rule: no matter how extreme the circumstances, no matter how ill or hopelessly injured potential donors are, they must die of some other cause before their organs can be harvested; it would never be acceptable to kill someone for his or her organs. But at the same time, ideally, a donor would be alive at the time his organs were harvested, because as soon as the flow of oxygenated blood stops, a process called warm ischemia quickly begins to ruin them. Early transplant medicine focused primarily on kidneys, which can be taken from a live donor and which survive ischemia passably well. But as surgeons began to focus their ambitions on livers, hearts, and lungs, they more and more faced the paradox created by their need for both a living body and a dead donor.

As it happened, doctors were also struggling with questions posed by another technology that had blurred the line between life and death: the mechanical ventilator. Artificial respiration dates back at least to biblical times, when, according to the second Book of Kings, the prophet Elisha (imitating, perhaps, Yahweh's puff of life into the nostrils of the man he had just created from dust) placed his mouth on the mouth of an apparently dead boy until "the child sneezed and opened his eyes." Further developments in reviving the near-dead were variations of the mouth-to-mouth technique (with the notable exception of the Native American practice of blowing tobacco smoke into the rectum of unresponsive people). But by the sixteenth century, doctors had supplemented this method with intermediary devices. In one of the public displays of dissection that he liked to arrange, the Renaissance anatomist and showman Andreas Vesalius cut a hole in a dog's throat, placed a straw in it, and breathed into the dog's lungs to demonstrate the doctor's ability to restore life. And in 1530, Paracelsus, a Swiss alchemist and physician, grabbed a bellows off the fireside of a patient who was not breathing and put the business end into the patient's nose. His attempt at artificial resuscitation was foiled by cinders, but eventually, inventors managed to perfect Paracelsus's idea. With the support of the lifesaving associations that sprang up in coastal cities and around the levees and canals of Europe, they developed nasal tubes and two-way pumps and even a manual ventilator that could fit into a doctor's pocket. Over the next three centuries, bellowslike devices, some of them strategically stationed near common drowning sites, greatly improved the prospects for the drowned and the asphyxiated.

But these machines had to be pumped by hand, ten or twenty times a minute, and patients who didn't respond quickly were soon given up for dead. The advent of electricity made a hands-free device possible, and by the late 1920s, a Harvard engineer funded by the Consolidated Gas Company (a company with an obvious interest in resuscitation for both its customers and its workers) figured out how to build a breathing machine that could be used over the long haul.

Like most inventors, Philip Drinker didn't originate the idea that made his fame and fortune. Others had thought of placing a patient in an airtight compartment from the neck down and using variations in air pressure inside the chamber to alternately compress and expand the chest, thus inducing something like the inhalation and exhalation of room air. But the Drinker Respirator, patented in 1928 and better known as the iron lung, bore the mark of modern scientific engineering as did none of its predecessors. Powered by a quarter-horsepower electric motor, it was self-sustaining and efficient, temperature- and humidity-controlled, and ready to be mass-produced. The apparatus looked like a hot water tank laid on its side. It had glass viewports, sealed inlets for reaching in with thermometers and other probes, and even an aluminum window through which X-rays could be taken. Drinker figured out how big to make the opening for the patient's head from information he obtained from the Knox Hat Company, and the selection of rubber collars that came with the machine were sized according to the data that had been amassed by the manufacturer of Arrow shirt collars. Drinker paralyzed cats with curare to test the machine's ability to induce breathing and persuaded Harvard lab workers to climb inside the iron lung and let him fiddle with the controls, taking charge of their breathing while measuring their air intake. He got Consolidated Gas's rescue crew to demonstrate their resuscitation technique so that he could compare its effectiveness to the machine's. Eventually, Drinker even described how doctors could improvise an emergency iron lung using inner tubes, shoe leather, and a vacuum cleaner.

Consolidated Gas found Drinker's work promising enough to buy a unit and donate it to Bellevue Hospital, where it was soon used to revive a victim of a drug overdose. But the first iron lung patient was an eight-year-old girl whose polio had paralyzed her chest muscles, a common effect of that scourge. She died of pneumonia after five days in the iron lung, but not before Drinker had proved in principle that a person could live in it long enough to survive the illness. Drinker's device—and its successor, the positive-pressure ventilator, which used modern electronics to deliver precisely measured quantities of air through nasal or throat tubes—soon led to the development of the intensive care unit, where machines take over the function of the brain stem. And although polio disappeared from hospitals, modern life, with its gunshot wounds and car wrecks and hypertension, supplied plenty of other patients with injuries that compromised the brain's ability to regulate breathing.

But it's easier to invent a machine that restores breath and life than it is to figure out when to use it—or, more precisely, when not to use it. As the use of artificial life support increased, so did the numbers of people who didn't improve, who lingered, unable to breathe on their own, inert and unresponsive even to the most noxious stimulus, and without any detectable brain activity or reasonable prospect of recovery. In 1959, two French doctors wrote up an account of twenty-three such patients. They were, the doctors said, in a *coma dépassé,* a state of total and permanent unconsciousness. These people would live until their hearts gave out, often a matter of hours, but sometimes of days or even weeks. (The heartbeat is not triggered by the brain stem.) Of course, there was no guarantee even that this would happen, no theoretical reason why they could not last indefinitely.

If the French doctors wondered whether removing the machines would be murder or mercy killing or simply a matter of letting nature take its course, they did not say. But in 1967, Henry K. Beecher, a prominent Harvard anesthesiologist, began to speak the unspeakable. "The time has come," he said "for a further consideration of the definition of death." Beecher explicitly connected this need with the "patients stacked up waiting for suitable donors" of organs. "Can society afford," he asked in a 1967 speech, "to discard the tissues and organs of the hopelessly unconscious patient when they could be used to restore the otherwise hopelessly ill, but still salvageable individual?"

His Yankee sensibilities as inflamed as his healer's, Beecher asked the dean of the medical school to form a committee to explore the issues of artificial life support and organ donation, which he believed were related. The Harvard committee, which Beecher chaired, included ten physicians, a lawyer, and a historian, and its report was published the following year in the *Journal of the American Medical Association.* "Responsible medical opinion," it announced, "is ready to adopt new criteria for pronouncing death to have occurred in an individual sustaining irreversible coma as a result of permanent brain damage." Heartbeat or no, the committee declared, patients whose brains no longer functioned and who had no prospect of recovering were not lingering but were already dead—brain-dead.

This physician-assisted redefinition of death gave transplant surgeons precisely what they needed: a class of people whose hearts were still beating but who were not alive, and from whom surgeons could harvest organs without being guilty of vivisection—but only if brain death could be made the law of the land. In the decade following the Harvard committee's pronouncement, however, and despite intensive lobbying by the American Medical Association, only twenty-seven states adopted brain death as a legal definition of death. Theoretically, this meant that someone who had been declared dead in North Carolina could be resurrected by transferring him to a hospital in South Carolina. Practically, it meant that a doctor procuring organs from a brain-dead person was not equally protected in all jurisdictions from the charge that he was killing his patient.

In 1980, a commission appointed by President Carter to look at bioethical questions in general took up its first cause: developing a model for state laws defining death as the irreversible loss of cerebral function. The commission had to give state legislators a way to convince their constituents that brain death was no mere legal fiction, and to do this they had to grapple with a question the Harvard committee had left unanswered—why, besides the fact that responsible medical opinion said so, the death of the brain is the death of the person.

Two rationales were considered. In one, called the "higher-brain" formulation, a brain-dead person is alleged to be dead because his neocortex, the seat of consciousness, has been destroyed. He has thus lost the ability to think and feel—the capacity for personhood—that makes us who we are and our lives worth living. But such quality of life criteria, the commission noted, raised uncomfortable ethical and political questions about the treatment of senile patients and how society valued the lives of the mentally impaired.

So the commission chose to rely on what it called the "whole-brain" formulation. The brain, it was argued, directed and gave order and purpose to the different mechanical functions of our bodies. If both the neocortex and the brain stem stopped working, a person could be pronounced dead, not only because consciousness has disappeared but because without the brain, nothing connects: there is no internal harmony, and the body no longer exists as an integrated whole. In addition, the president's commission said, the traditional criteria of death—the cessation of breathing and pulse—had all along been secondary, that thanks to increased medical knowledge, it was now clear that these were only signs that the brain had died. In other words, brain death was not a mere rejiggering of death to suit the needs of transplant surgeons and their patients. It was death new and improved. The president's commission's strategy paid off. By 1990, brain death was the law of the land in all fifty states.

EVEN AMONG MEDICINE'S NOBLE LIES, brain death stands out, and not only because it blurs a distinction that most of us think of as absolute or because it moves people who are warm and flush and breathing onto the other side of the divide between life and death or because the precise means of papering over the underlying fractures were spelled out so plainly (and by a presidential commission, no less). Brain death is unique because it was invented for the benefit of someone other than the patient. The noble lies we've looked at so far were created in order to allow doctors to help the patient himself or herself. In the case of brain-devastated people, it would be sufficient to establish a means by which life support can be discontinued, which a series of court cases—those of Karen Quinlan, Nancy Cruzan, and Terri Schiavo, for example—and policy refinements have done. You don't need to call people dead in order to justify turning off their machines. The only reason brain death was created was to make possible what would otherwise be unthinkable: turning a person into a repository of organs, treating a human being as pure means.

Medicine's noble lies perhaps always have this function: rendering the impermissible possible and providing an end run around prohibitions that would be too hard to clear out of the way by another method. Turning addiction into a chronic disease creates the opening for forced abstinence in a society that has rejected the political version of Prohibition. The diagnosis of depression makes it possible for people who are suffering to get access to and take consciousness-altering drugs in a society that frowns on doing that. The rendering of homosexuality as an innate and immutable biological condition makes it permissible for people to conduct their sex lives in a way that other people find repugnant. Calling Ted Kaczynski crazy allows us to ignore the truth of his critique of technological society. And in each case, the diagnosis helps us to avoid a confrontation with an unsettling reality, a flaw in our social fabric, a conflict in our moral code.

That confrontation is precisely what happens when science can't fabricate a diagnosis that satisfies all of the parties to a conflict. Abortion, of course, is the paradigm case. Like brain death, abortion entails the ending of a life for someone else's benefit. But the apparent costs and benefits of abortion are not so stark as they are in brain death. The liberty of a pregnant woman is a less compelling good than is the imminent death of a person with kidney failure, and a fetus does not strike everyone as a full-fledged human being worthy of the protection of the law. Medical science is no help here. If anything, it has confused the picture, blurring the traditional distinction between the pre- and postquickening fetus with its ever-shifting definitions of "viability," its always-increasing ability to keep premature infants alive. So we are left with one faction attempting to keep the attention on the issue of "choice," as if a life weren't at stake, and the other on the issue of "life," as if liberty didn't matter.

The abortion debate keeps squarely in front of us the fact that two of our most cherished goods—life and liberty—are not always reconcilable. Sometimes we have to decide between the two. In fact, decisions like this get made all the time. Just consider how precipitously the automobile fatality rate (currently more than forty thousand deaths per year, with nearly half of those due to drunk driving) would drop with the introduction of surveillance devices—engine-mounted speed and g-force monitors that transmit a record of your driving to the authorities, for instance, or Breathalyzers that make it impossible for intoxicated people to start their cars. Now imagine the outcry that would occur were these devices, both currently available, to be made mandatory for all drivers. (Indeed, a 2008 Alaska legislative proposal requiring Breathalzyer-based ignition interlocks after a single drunk driving conviction prompted the American Beverage Institute to complain that "moderate drinkers" didn't deserve to be "saddled" with this inconvenience and to prophesy that the time when this "silliness" would spread to every car in every state was "closer than you think.") In this case, freedom (or maybe it's the pursuit of happiness on the open road) trumps life, but it happens so deep in the background that we are not aware of either the conflict or its resolution.

Capitalism is full of "decisions" like this—and it may be that the notion that the invisible hand will guide us to progress is its own kind of noble lie—but the free market is not going to end the abortion debate any sooner than medicine or science is. Indeed, it is possible to glimpse in the organized opposition to abortion the deep discomfort that the irreconcilability of life and liberty stirs, at least among a sizable segment of the population. While it would be a mistake to think that renewed moral debate will always end in interminable wrangling or, worse, in the imposition of sanctions like the Bush administration's prohibition of funding for stem-cell research, the story of brain death is indicative of what would happen in a fractured society without our noble lies. Indeed, the success of brain death seems nearly miraculous in retrospect. Were a university committee to propose such an idea today (assuming a group of professors could agree on such weighty matters), it is hard to imagine that it would become the law of the land as easily as it did.

WHEN NICK BREACH DECIDED to become an organ donor, Children's Hospital directed the family to Gift of Life, an organ procurement organization based in Philadelphia. With a staff of a hundred, the agency is among the largest OPOs in the nation. In 2007, it helped to manage over a thousand organ donations at more than 150 member hospitals in Pennsylvania, New Jersey, and Delaware—nearly 5 percent of the total organs removed in the country. Inside its offices in downtown Philadelphia, phones ring, intercoms squawk, and well-dressed people scoot intently through hallways. Were it not for the snatches of conversation drifting up from the cubicles—"We've got a coma in Pottstown"; "They're fighting about the other half of the liver"—and the whiteboards tracking the progress toward death of patients in area hospitals, you'd think you were in the headquarters of a financial services corporation.

That's not an accident. "We were the first OPO with a business plan," Howard Nathan, the CEO of Gift of Life, told me. He's been here since 1978, when he got a job as a transplant coordinator, and since he took over in 1984, he has worked to bring established business practices into the agency. When you get put on hold at Gift of Life (1-800-KIDNEY1), you hear not soft rock but hard-sell messages from donor families and grateful recipients. On the tape, one donor family says, "I made the decision to make a situation that was very wrong right."

Gift of Life isn't selling organs, of course. The phone ads, the pamphlet featuring "A Message from Michael Jordan," the bumper stickers that say, "Don't take your organs to heaven—Heaven knows we need them here" are all promoting an attitude about how, as Nathan put it, "society should feel about this subject." Gift of Life, according to its mission statement, wants to "positively predispose all members of the community to organ and tissue donation so that donation is viewed as a fundamental human responsibility." That's why, according to Kevin Sparkman, its director of public relations at the time (and now of community relations), the agency aggressively publicized Nick Breach's decision. "Here's a young man who is awake and aware, contemplating his death, and he becomes a donor," explained Sparkman. "What a great example of what we want families to do!"

Nathan, who is fifty-one, has never worked anywhere other than Gift of Life, and he seems to hold himself personally responsible for every person on the organ waiting list. He's willing to push the envelope, proposing, for instance, a state law that provides burial benefits to donors, a measure that some said would commercialize donation. In 1994, he advocated for a federal law requiring all hospitals to report all deaths and imminent deaths to their regional OPO so that it could determine which cases might be suitable sources of organs. This last measure has meant a more than tenfold increase in referrals to his agency, which now sorts through between three and four thousand calls every month, winnowing out those who are too old or sick, who have been dead for too long, or who will die the wrong way; and then trying to obtain family consent from those who are eligible to donate—a process that it (and all OPOs) goes through even if the dying person has a valid organ donor card. ("Dead people can't sue," explained Nathan, adding that the real problem is that you can't risk the headline "Organs Donated against Family Wishes.") Gift of Life's procurements have increased by more than 60 percent since 1994, according to Nathan, precisely what the business plan predicted.

Nathan knows the idea that the total destruction of the brain is the death of the person is crucial to his agency's success. "Organ donation is based on public trust," he told me. "The dead donor rule is necessary for public trust, so it's essential to what I do. Without it, I'd have to get another job." That's why so much of Gift of Life's work is educating people about brain death.

Brain death protocols at most hospitals include the directive to notify the regional OPO. When Gift of Life gets the call, it dispatches a transplant coordinator to the hospital to try to obtain the family's consent—an essential step because an organ donor card is not a legally binding document; it is up to the family to decide whether to allow harvesting, and family members are legally free to override the patient's wishes.

"The first thing we do is ensure that the family understands and acknowledges that their loved one is dead," Linda Herzog, a senior hospital-services coordinator, told me. Families who think donation is actually going to kill the patient refuse more often, she said, than families who think their relative is already dead—a finding that is not nearly as surprising as the fact that a substantial minority of families give consent even though they remain convinced that their brain-dead loved one is actually alive.

Bodies on respirators have always confused people. When the corpse of a drowned sailor was used to demonstrate a bellows resuscitator in 1840, according to a doctor who was present, "the body was made to breathe in such a manner as to lead the bystanders to suppose that the unfortunate individual was restored to life." The inventor had to disabuse his audience of this notion, and it falls to Herzog to do the same for families of what the transplant industry calls "heart-beating cadavers." For this reason, Gift of Life has developed a program that trains hospital staffs to explain to family members why the person whose chest is rising and falling, who is flush and warm to the touch, is actually dead. I watched in a darkened conference room as Herzog reviewed the program for two transplant coordinators who were scheduled to present it later that afternoon in a Philadelphia hospital.

Using slides, Herzog ran through the process by which brain death is established. A neurologist performs a series of tests at the bedside, checking for such things as pupillary reflexes, response to pain, and the ability to breathe spontaneously. (If the patient is entirely unresponsive during two such examinations, the doctor concludes that his or her whole brain—cortex and brain stem—has been destroyed.) This is not a terribly sophisticated procedure, but it's far more complicated than, say, ascertaining that a person has no pulse, and far less self-evident. Even when the tests are conducted or reenacted in front of family members, they often rely on their intuition and insist that the patient is still alive—a confusion compounded, according to Herzog, by the fact that in the intensive care unit, families tend to pay more attention to the monitors than to the patients, and the monitors continue to register vital signs even after brain death has been declared. (The body must be kept stable until organs are harvested.) This failure to accept the truth is a function of denial, Herzog said, and she went on to note, with some dismay, that even highly trained professionals who fully accept the concept sometimes talk to brain-dead patients.

"It took us years to get the public to understand what brain death was," Nathan said. "We had to train people in how to talk about it. Not that they're brain-dead, but they're dead: 'What you see is the machine artificially keeping the body alive . . .'" He stopped and pointed to my notebook. "No, don't even use that. Say 'keeping the organs functioning.'"

Virtually every expert I spoke with about brain death was tripped up by its semantic trickiness. "Even *I* get this wrong," said one physician and bioethicist who has written extensively on the subject, after making a similar slip. Stuart Youngner, the director of the Center for Biomedical Ethics at Case Western Reserve University, thinks that the need for linguistic vigilance indicates a problem with the concept itself. "The organ procurement people and transplant activists say you've got to stop saying things like that because that promulgates the idea that the patients are not really dead. The language is a symptom not of stupidity but of how people experience these 'dead' people—as not exactly dead." Brain death, it seems, is an epistemological hybrid: part fact, confirmed by science, and part philosophy, confirmed by faith alone.

A BOY NOT MUCH OLDER than Nick Breach is central to one doctor's heresy against that faith. Alan Shewmon, a professor of pediatric neurology at UCLA, says that he has evidence that brain death is not the death of the person, and he's showing it right now at the Third Conference on Coma and Death in Havana. Playing in the corner of this darkened and chilled room is a video that Shewmon calls *Seeing Is Believing*. In it, a boy is recumbent on a bed, his feet toward the camera, and his legs bowed, almost froglike. He's wearing shorts but no shirt, and there are two tubes entering his body, one in his abdomen, the other in his throat. The boy's chest rises and falls to the whir and click of the respirator, but otherwise he is perfectly still.

On the tape, Shewmon stands near the bed conducting a medical examination. He looks into the boy's eyes, shakes maracas next to his head, inserts a swab in a nostril, drops cold water into his ears and lemon juice on his tongue, pinches and palpates and inspects. None of these actions draws the slightest response from the boy— the expected result in a boy who, Shewmon has told us, meets all the clinical criteria for brain death. But he's not dead, not legally anyway. The boy, in fact, is at his home, cared for by his mother.

Havana is the first stop on a five-city world tour that will take Shewmon as far as Japan. By day, he will show this video and will lecture on its implications for the concept of brain death. By night, he will give a piano recital whose program includes Franz Liszt's *Pensées des Morts*. Shewmon, who is trim and balding and always dressed in a suit and tie, has been thinking about death for most of his career. A practicing Catholic, he has made contesting the concept of brain death a specialty and has served on the Pontifical Academy of Sciences task force on the subject. Shewmon's inquiry has led him from the higher-brain rationale through the whole-brain rationale to his current position: a strong conviction that brain death, while a severe disability, even severe enough to warrant discontinuing life support, is not truly death. His position is well known—and scandalous—among the neurologists, bioethicists, and anthropologists gathered here. To the extent that there can be a buzz at a medical conference, it's been about Shewmon and his video (this is its first screening for a semipublic audience), and the fact that he seems to be accompanied everywhere by a small coterie of young, fully frocked priests has not diminished his infamy here.

Although the boy on the video, whom Shewmon calls Matthew, doesn't seem dead, it is hard to think of him as alive. A nurse removes the upper tube and suctions the small breathing hole in the boy's throat; on the video, Shewmon notes out loud that the boy did not cough and continues the exam. Then something different happens: some ice water trickles onto the boy's shoulder, and it twitches. And although the screen is too small for the audience to see this, Shewmon tells us that at this moment Matthew is sprouting goose bumps, that his flesh is mottling and flushing with the stress of the exam. He is showing signs, that is, of precisely the kind of systemic functioning that the brain-dead are not expected to have.

In the video, Shewmon lifts Matthew's arm by the wrist, and the hand springs to life with a small spasm. A woman's voice—Matthew's mother, we soon learn—says, "When he knows what you're going to do, he stops that." Shewmon lets this implication of the boy's agency hang briefly in the air before he describes for us what is going on in medical terms: clonus, an involuntary contraction and release of nerves. He is making his main point: that this boy—who at age four was struck with meningitis that swelled his brain and split his skull, who would probably have been pronounced brain-dead had he not been too young under the statutes of the time, whose mother refused to discontinue life support and ultimately took her son home on a ventilator and a feeding tube, who had persisted in this twilight condition for thirteen years, healing from wounds and illness, digesting his food, metabolizing, and growing—is alive. Not by virtue of intention or will, but because he has maintained a somatic integrated unity—the internal harmony, and the overarching coordination of his body's functions—which, if the whole-brain rationale is correct, he simply should not be able to do.

After the presentation ends, I speak to Ronald Cranford, a professor of neurology and bioethics at the University of Minnesota, who is one of Shewmon's most outspoken critics. He tells me that

Matthew's case is only an unusually prolonged example of the normal course that brain death takes. "I agree that the child is brain-dead," he says. "But you have to understand that any patient you keep alive, or dead, longer than a few days will develop spinal cord reflexes." Cranford recalls a case in which the doctor said, "Yes, she's been getting better ever since she died."

That night, the conference attendees are bussed to an ornate concert hall in central Havana for Shewmon's recital, and on the return trip I snag the seat next to his. As we wend our way through the old city, I ask him what moral he wants his audience to draw from his presentation earlier in the day. Shewmon, who has authored articles in Catholic journals about the metaphysics of the soul and written effusively about his own Chopin-inspired encounters with the holy, denies having any such thing in mind. "The ethicists wanted to immediately draw me out on the moral implications, but that's not the point. The video may have moral implications, but let's just understand the facts. I'm showing something that's very iconoclastic at an empirical level." It is, he tells me, the doctors who believe that brain death is the death of the person who are letting morality contaminate their work. They are making the momentous judgment that people like Matthew are so devastated that they have lost their claim to existence. Shewmon promises to lay bare the nonscientific nature of this claim at his talk the next day.

He also promises that his lecture—a comparison between the brain-dead and people who are conscious but have been paralyzed by injuries to the upper spinal cord—will be "agonizingly detailed," and he does not disappoint. After he makes his case that the only significant medical difference between someone like Christopher Reeve and someone like Matthew is that Reeve retained consciousness (and, of course, couldn't be pronounced dead) and that in both cases the body was disconnected from the brain, no one really takes issue with his science. At the same time, none of the physicians will accept what Shewmon is really saying: that the brain-dead are not dead. "The main philosophical question is, Is this a body or is this a person?" says Calixto Machado, the

Cuban neurologist who organized the symposium. Fred Plum, the chairman emeritus of the Department of Neurology at Cornell University's Weill Medical College, has positioned himself directly in front of the podium for the talk and shoots his hand in the air as soon as Shewmon is finished. "This is anti-Darwinism," Plum says. "The brain is the person, the evolved person, not the machine person. Consciousness is the ultimate. We are not one living cell. We are the evolution of a very large group of systems into the awareness of self and the environment, and that is the production of the civilization in which any of us lives."

Shewmon has sprung his trap. He hopes to break down the pretense that anyone subscribes to the whole-brain rationale. He wants to show that the higher-brain rationale—which holds that living without consciousness is not really living, and which the president's commission rejected because it raised questions about quality of life that science can never settle—is the sub-rosa justification for deciding to call a brain-dead person dead. He looks past Plum and says with his usual mildness, "I interpret Dr. Plum's point as agreeing that we have moved from talking about the biological organism to the personhood/consciousness rationale. That's precisely my point as well. So thank you."

BRAIN DEATH MAY BE NO WORSE than any other creation of a committee. It's just that the stakes are unusually high, so it is disconcerting to discover how hard it is to find a neurologist who actually subscribes to the whole-brain rationale, and, when you find him, how vague his evidence is—and how circular his reasoning. The neurologist James Bernat, a professor at Dartmouth Medical School and the author of the chapters on brain death in several neurology textbooks, is among the foremost defenders of the whole-brain concept. Like Shewmon, Bernat served on the Pontifical Academy of Sciences task force. And in August 2000, his position appeared to prevail when Pope John Paul II, speaking before an international transplantation congress, said that "the complete and irreversible cessation of all brain activity, if rigorously applied," along with the family's consent, gave a "moral right" to remove organs for transplant, thus resolving an ambiguity in the church as to whether Catholics should become donors. But even Bernat sees the problem he's up against. "Brain death was accepted before it was conceptually sound," he told me on the telephone from his office in New Hampshire. He readily admitted that no one has yet explained scientifically why the destruction of the brain is the death of the person rather than an extreme injury. "I'm being driven by an intuition that the brain-centered concept of death is sound," he said. "Death is a biological function. Death is an event."

Stuart Youngner, of Case Western, however, is not so certain. A white-bearded, avuncular professor of psychiatry, Youngner calls the Harvard committee's work "conceptual gerrymandering," a redrawing of the line between life and death that was determined by something other than science. Youngner takes a psychiatric approach to brain death, railing against what he sees as the bad faith behind the idea and urging that we confront its ambiguities and the ambivalence that it evokes. "What if the Harvard committee, instead of saying, 'Let's call them dead,' had said, 'Let's have a discussion in our society about whether there are circumstances in which people's organs can be taken without sacrificing freedom, without harming people'?" Youngner wondered. "Would it be better?"

The problem, as Youngner sees it, is that the veneer of scientific truth lies over the concept of brain death, concealing the troubling fact that the lives of brain-dead people have ended only by virtue of what amounts to a social agreement. According to Youngner, this means that the brain-dead are really just "as good as dead," but, he is quick to add, this doesn't mean that they shouldn't be organ donors. Instead, he suggests that "as good as dead" be recognized as a special status, one that many people, brain-dead or not, may achieve at the end of life. "I'm willing to point out the ambiguities and inconsistencies in the notion, and I actually think that acknowledging them may in the long run be better," he told me.

During the last decade, Youngner and other doctors and ethicists have developed protocols to allow critically ill or injured people who have no hope of recovery, but who are unlikely to become brain-dead, to donate their organs after they have been declared dead by the traditional cardiopulmonary criteria. This procedure, which is known as non-heart-beating-cadaver donation and which requires extremely rapid intervention and newly developed preservation techniques, may make it possible to salvage viable organs in a wider range of cases.

Because Nick Breach was so unlikely to suffer brain death, he was a likely candidate for this procedure. This was the part that had his parents the most confused—that their best hope for fulfilling his last wish lay in getting him to a hospital before he stopped breathing on his own, placing him on a respirator, and saying their good-byes before he was wheeled, still alive, into an operating room. There his death would be tightly choreographed: doctors would remove his ventilator and wait for his heart to stop. If that took more than an hour, warm ischemia would set in (as his breathing would be too compromised to supply oxygen to his organs), the donation would be aborted, and Nick would be returned to a hospital room to die. But if cardiac arrest came in time, a five-minute count would begin, at the end of which Nick would be declared dead. A transplant team standing by in an anteroom would immediately harvest his organs and rush them to their recipients. (Even with this alternative, the window for success was fairly narrow. "All we're trying to do," Howard Nathan acknowledged, "is give it a shot.")

Non-heart-beating protocols, according to the Institute of Medicine, have the potential to increase donation by as much as 25 percent. But, as Youngner points out, doctors could attempt to revive the donor patient at any time during the waiting period; a closely managed death is irreversible only because everyone has agreed that these measures aren't going to be taken. So, says Youngner, the five-minute waiting period (or two minutes or ten minutes, as other protocols would have it) is really just a decent interval, a more or less arbitrary marker of the passage from life to death, whose significance is far more symbolic than scientific.

Robert Truog, a professor of medical ethics and anesthesiology at Harvard Medical School, is even more critical of the protocol. "Non-heart-beating protocols are a dance we do so that people can comply with the dead-donor rule," he told me. "It seems silly that we hang on to this facade. It's a bizarre way of practice, to be unwilling to say what you are doing"—that is, identifying a person as an organ donor when he is still alive and then declaring him dead by a process tailored to keep up appearances and which, in the bargain, might not best meet the requirements of transplant. In Truog's view, a better approach would be to remove these patients' organs while they are still on life support, as is done with brain-dead donors. "If they have detectable brain activity, then they should be given anesthetic," he said, but there is no reason to continue to conceal what is happening by waiting for their hearts to stop beating.

Abandoning the dead-donor convention, which is an inevitable consequence of Youngner's and Truog's positions, may, however, cause other problems. It awakens the same sort of fears that Christiaan Barnard was trying to forestall at the inception of heart transplant: that doctors would, as Truog put it, "seem like a bunch of vultures." Without the rule, Truog pointed out, "taking organs is a form of killing"—killing that he thinks is justified, and that Youngner and others would argue is already happening even in the brain-dead.

In return, Truog pointed out, patients would gain more control over the end of their lives: they would no longer have to wait until they crossed over that gerrymandered border and instead could specify at what point they would like to be declared dead enough to donate their organs. This, however, might not be adequate consolation for those who fear that the need for organs might create a perverse incentive for doctors to give up on them, after weighing their lives against those of others who may be more worthy or less damaged. Youngner expressed reservations about how his position would sound to other doctors and, most important, to potential donors.

"I think that stridently advocating the abandonment of the dead-donor rule would be a mistake," he said. He worried, he told me, that religious conservatives and others might "seize on it as a violation of the right to life," thus turning transplant into another medical practice, like abortion or fetal stem-cell research, that's bogged down in intractable political wrangling.

AFTER NICK BREACH MADE HIS DECISION, one of his first questions was whether he would be dead when his organs were taken. His parents told him that he would be, and, in a way, they saw this as one of the few things they could be sure about. Rick and Kim were more troubled by their son's next concern—that he might be taken from them prematurely. Thus began a vigil that took on a strange dual nature: keeping Nick company, making him comfortable, spending as much time as possible with him, and, at the same time, monitoring him for the signs that death had come so close that it was time to get him to the hospital so that he could become an organ donor.

The Breaches had to rely on the doctors to know what those signs were, just as they would have to take the doctors' word that Nick was really dead when his organs were harvested. But no one told them about Alan Shewmon or Henry Beecher, and they might have been discomfited to know that doctors don't necessarily agree about when a person is dead. And they certainly weren't aware—who is?—that this disagreement is as old as modern medicine. Long before transplant was a possibility, scientists have known that, as the eighteenth-century Danish anatomist Jacques-Begnine Winslow put it, "Death is certain, since it is inevitable, but also uncertain, since its diagnosis is sometimes fallible." There has always been a great deal of moral and conceptual ground between the quick and the dead, and if science has long been preoccupied with discovering the boundary, it has never been particularly successful in finding one that provides both certainty and infallibility.

This may be why, until modern times, doctors largely avoided the question of determining exactly when someone is dead. Ancient physicians considered the determination of death outside the scope of their job, which was to inform the family that death was imminent and then to withdraw from the case. Death was declared by the family and certified and tallied by bureaucrats, and the actual moment of its occurrence was left indistinct. Doctors got involved after the Enlightenment, when the moment of death became yet another of those mysteries that science undertook to clear up.

This might have been a purely academic exercise, just another opportunity for scientists to use the power of reason to pry loose nature's secrets. But it took on urgency in the mid-eighteenth century, when *On the Uncertainty of the Signs of Death,* Winslow's treatise claiming that doctors hadn't really located the boundary between life and death, was translated into French by Jean-Jacques Bruhier. Bruhier, an ambitious Parisian doctor, added something to the text: 181 case studies of people who he claimed had been declared dead prematurely. He spared no macabre detail in these accounts: a woman whose corpse had been found on the stairs of her crypt, her fingers gnawed off in an apparent attempt at nourishing herself; a French aristocrat revived in his coffin when a doctor blew pepper into his nose; a woman who woke up shrieking when the bed in which she was laid out caught fire. Bruhier wanted people to know that the absence of pulse and respiration was not the sure sign of death that doctors said it was and that people were thus in imminent danger of being buried alive. He took a fundamentalist approach, arguing that putrefaction was the only sure sign of death.

Bruhier didn't originate the fear of premature burial. Legend has it that Andreas Vesalius was hauled before the Spanish Inquisition for having cut open a body in which the heart was still beating, and that he was ordered to make a pilgrimage to Jerusalem in atonement—a journey on which he died. Daniel Defoe, in his *Journal of the Plague Year,* tells the story of a piper who passed out drunk on the road and, when he came to, found himself rattling along on a death cart, buried in a heap of bodies destined for a mass grave. ("Hey, where am I?" he yelled out, according to Defoe's fictional narrator. "Why, you are in the dead cart," came the reply, to which the piper responded, "But I ain't dead, though, am I?") And, according to Jan Bondesen's *Buried Alive: The Terrifying History of Our Most Primal Fear,* in the 1720s, the French drove pins into the toes of plague victims to make sure that they were really dead before burying them.

Winslow's book was translated into English, Swedish, Italian, and German (with each translator adding his own stories about people committed to the grave before their time), sowing seeds of doubt across Europe about whether the dead were really dead. Other developments, particularly those in artificial respiration, further confused the question of when people could reliably be declared dead, and by the early nineteenth century the anxiety about premature burial was widespread enough to lead some people to write wills ordering their hearts to be removed prior to interment. The fear of being buried alive also led communities to build leichenhausen—waiting mortuaries, where bodies could be laid out until they either stirred (watchmen were posted to monitor the beds) or rotted. And then there were the safety coffins, rigged with bells to signal revival and inlets to provide air until someone could dig up the not-yet-dead body.

By the turn of the twentieth century, the premature burial panic had passed, most likely because no one in the safety coffins or waiting mortuaries ever actually woke up. According to George Behlmer, a scholar of the Victorian era, the anxiety had never been warranted by events. Rather, he said, it was a version of the moral panic that accompanied the scientific revolution and its blasting away of familiar ethical benchmarks. In this case, the public feared turning over this momentous event to medical experts, with their stethoscopes and arcane language and specialized knowledge, not to mention their well-known interest in opening up dead bodies to see what was inside. (Indeed, while there may not have been many premature burials, there were certainly grave robbers commissioned by anatomists to bring them bodies for autopsy.)

The Breaches seemed not only unafraid but relieved and willing to trust the doctors. Yet if they had known how ambiguous the determination of brain death is, and the confusions that can result, they might not have been so trusting. When a doctor determines brain death, according to the practice guidelines of the American Academy of Neurology, he or she can't just order some lab tests to determine, say, whether there is any electrical activity in the brain. In fact, electroencephalograms and ultrasounds are an optional part of a routine that, much like Shewmon's with Matthew, relies on old-fashioned, hands-on doctoring, in which evidence of the brain's total destruction is indirectly assayed by methods that include the ancient Marseillesian technique of pricking the nail bed to see whether the patient responds in any way to pain. Brain death is a "clinical diagnosis," say the guidelines, which means that doctors must learn from experience to distinguish between, for instance, movements of the limbs that indicate ongoing brain activity and those that are merely autonomic reflexes. And as with any diagnosis that depends on a physician's judgment, it is possible to make mistakes.

Gail Van Norman, who teaches and practices in Seattle, has collected stories of these mistakes. For example, she wrote, an anesthesiologist determined that a woman already pronounced dead and prepared for organ removal was showing signs of life. The doctor administered some medications that elicited even stronger responses. Organ collection was canceled, and the patient eventually went home, neurologically impaired but decidedly alive.

Two of Van Norman's other stories didn't have such happy outcomes. In the first, the donor began to breathe spontaneously during the harvest. The attending anesthesiologist, according to Van Norman, then reviewed that patient's chart and found that after the ventilator was removed (part of the evaluation is determining whether patients can breathe on their own; if they can, then they can't be brain-dead), the patient had begun to breathe.

The neurosurgeon pronounced death anyway, and the patient was prepared for surgery. The discovery that he was still alive came too late to save him, as his liver had already been removed.

In the second case, the donor also began breathing on her own in the operating room. An anesthesiologist tried to stop the proceedings, but the doctor who had declared her dead noted that the liver recipient was in imminent danger of dying and that the donor's chances for meaningful recovery were in any event nil. They removed her liver, and, needless to say, she died. The story has a terrible coda: the recipient died of an acute hemorrhage before liver collection was completed. No other recipient could be found in time, and the liver went untransplanted.

From these stories, Van Norman concludes that her fellow anesthesiologists must be prepared to rein in the occasional cowboy surgeon in order to maintain the clear line between life and death. She recognizes how tempting it is, in the face of people desperate for organs, to blur that line and how easy it is for a well-intentioned doctor to move a boundary that was arbitrarily determined in the first place. And she is clear about what is at stake—not simply the dignity of the dying but the public trust that is ensured by the dead-donor rule.

AS NICK BREACH THOUGHT about his death, he made some additional last wishes that were, thankfully, easier to satisfy than his desire to become a donor: Ronald McDonald came to visit; so did Weird Al Yankovic, one of Nick's idols. When Yankovic pulled up to the Breaches' house in a bus, the neighbors moved their cars to accommodate him. Yankovic came inside and sat for a while with Nick, who was bedridden by then. Nick told him, "I really love all your CDs, Weird Al."

Six days later, at 11:45 P.M., Nick stopped breathing. Rick, who was taking his turn by the bedside, summoned Kim, who called an ambulance and began to administer CPR. The plan was to revive Nick so that he could be brought to a hospital and placed on a ventilator. But his mother's efforts, and those of the paramedics in the ambulance and the staff in the emergency room, failed. Nick's heart had stopped too soon, and ischemia had set in. In the end, the only organs he was able to donate were his eyes.

It is tempting to wish that death weren't so complicated. Had Nick and his parents realized how alive he still needed to be in order to donate his vital organs successfully, they could have been given an honest choice between having Nick remain at home until the end and giving up on his goal of becoming a donor and going to the hospital much earlier and staying until he could be declared "as good as dead."

Over and over again at the conference in Havana, I heard ambivalence and anxiety about "the public" knowing what doctors already know. "These things ought to be worked out in the medical profession, to some extent, before you go to the public," Shewmon told me. "Because if you go public right away, it could just put the kibosh on the whole thing because people get hysterical and misunderstand things." He paused and looked at me. "These are complex issues. You can't expect the public to understand these things in sound bites, which is what they usually get. So I'm reluctant to talk to reporters about this stuff."

During a break between sessions, I got into a conversation with John Lizza, a philosopher who has written extensively on the subject of brain death. He told me that he had been talking about this subject with a colleague, and that they'd found themselves calling brain death a noble lie. Later, as the conference reconvened, I asked him whether we could talk some more about that idea. He was visibly upset. "Listen, I'm not sure about that comment," he said. "It's inflammatory. It's too strong." Among his concerns, he explained, was the possibility that his words might discourage people from becoming organ donors.

It may be too much to say that the concept of brain death is a noble lie, but it is certainly less than the truth. Like many of technology's sublime achievements, organ transplant, for all its promise, also has an unavoidable aspect of horror: the horror of rendering a human being into raw materials, of turning death into life, of harvesting organs from an undead boy. Should a practice, however noble, be able to hold truth hostage? Perhaps the medical profession should embrace the obvious: to be an organ donor is to choose a particular way to finish our dying, at the hands of a surgeon, after some uncertain border has been crossed—a line that will change with time and circumstance, and one that science will never be able to draw with precision.

6

Persistent Vegetative State: Back from the Dead

For someone left for dead thirteen years ago, Candice Ivey seems to be doing pretty well. She's still got her bright eyes and blond hair, her homecoming-queen good looks and A-student smarts. She's got a college degree and a job as a recreational therapist in a retirement community. She has, however, lost her ballerina grace; now she walks a bit as if her feet are asleep. She slurs her words a little, too, which sometimes leads to trouble. "One time I got pulled over," she tells me in her snappy Carolina twang. "The cop looked at me and said, 'What have you been drinking?' I said, 'Nothing.' He said, 'Get out here and walk the line.' I was staggering all over the place. He said, 'All right, blow into this.' Of course, I blew a zero and he had to let me go."

Candice doesn't remember anything from an hour or so before she pulled out of her high school parking lot and got t-boned by a logging truck until about two months afterward, a span that included Christmas and New Year's. But it's all seared into the memory of her mother, Elaine, especially the part about the doctors telling her that there was no hope for Candice. She was brain-dead, they said, and she would stop breathing shortly after they disconnected the respirator she'd been on for nearly a week. The organ procurement people visited, but Elaine had to tell them that Candice had expressed a desire not to be an organ donor. Her gurney was wheeled into a small room adjoining the intensive care unit. "I crawled into her bed, put my arms around her, and they pulled all the tubes out," Elaine told me. "That was the first night I actually slept, and the next morning she was still breathing."

No one knows why Candice didn't die, but Elaine Ivey didn't have time to think about this. She now had another terrible problem on her hands. "I thought, my child does not want to lay with the ventilator keeping her alive," she said. "That was the easy part. But when they tell you her brain is there, that she can breathe on her own, then you have a different kind of choice. You have to decide whether to feed your own child." The doctors discouraged her, saying that breathing might be about all that Candice would be able to do independently. But the doctors had already been wrong once, so Elaine decided to have the feeding tube reinserted. Candice's neurosurgeon was livid, Elaine told me. "He said, 'What the hell have you done?' He thought I was just prolonging her agony, and that I would have a vegetable on my hands. But when it's your child lying there, you'll do anything."

Anything included letting a doctor named Edwin Cooper, who approached Elaine out of the blue four days after the accident, try his experimental treatment on Candice. It was an electrified cuff on her wrist that sent a 20-milliamp charge—enough to make her hand clench and her arm tremble—into her median nerve, a main line to the brain. It might rouse her from her coma, he said.

"I thought it was hokey, if you want to know the truth," Elaine said. She agreed nonetheless, and the cuff went on. (She was, she said, "drunk as a coot" from a combination of "nerve pills and a full glass of whiskey.") Within another week or so, Elaine Ivey was sure that Candice was stirring. Her doctors doubted it. "They kept telling me it was just reflexes," Elaine said, "but a momma knows." And just before New Year's Day, a month after the accident, Cooper asked her how many little pigs there were. Candice held up three fingers.

Now twenty-nine, Candice Ivey is thrilled to see Cooper when he arrives at her door on a spring morning thirteen years later. She gives him a big, warm hug, sits close to him on her couch, and easily agrees to let him test her coordination for my benefit. They both come to tears when they reminisce about the time she recounted her ordeal to some nurses at Cooper's hospital. She's gracious when she shows me the photos of her and her mother and her identical twin sister, her collection of small books, her tidy condo, and she'll tell me anything I want to know about her brain injury and its aftermath. What she really wants to get across, however, and what she keeps coming back to, is her gratitude. "The wreck was my fault," she says. "But getting better, that was God's doing. He sent Dr. Cooper to my momma, didn't he?"

Edwin Cooper has been sent, or has sent himself, to about sixty severely brain-injured people and their families since 1992. While recoveries like Candice Ivey's are rare, he claims that he can use his electrical stimulator to awaken the sleeping brain and help comatose patients to recover function more quickly than if they were left untreated. He's run a couple of small experimental studies and published a few articles, but his therapy remains unproven and largely ignored in the United States. Some prominent doctors are quietly looking into a related treatment here, but his best hope for proof comes from Japan, where over the last two decades neurosurgeons have successfully used electrical stimulation to treat hundreds of patients, some of whom have been unconscious for many years. If research on either side of the Pacific confirms Cooper's claims, it may bring hope to the families of patients now considered beyond cure. But it may also undermine the hard-won but fragile consensus on what, neurologically speaking, makes life worth living and when it is acceptable to pull the plug.

SUSPENDED IN ITS CASE of hard bone, isolated from the outside world, its internal environment exquisitely modulated by the ebbs and flows of neurochemicals, your brain can withstand many insults before it just plain shuts down. But car wrecks are not among them, at least not the kind Candice Ivey had, in which g-forces slam your brain into the inside of your skull like a hurricane tossing a ship onto a reef. You might become immediately unconscious, a sign that the tendrils that carry signals from one neuron to another—the white matter—have been sheared by the twisting of your brain on its stem. Or you might arrive at the hospital lucid and awake, only to slowly lapse into unconsciousness as the bruised and lacerated tissue swells, the blood starts to pool, and, with nowhere else to go, the gray matter presses in on itself until the white matter is damaged or destroyed. And then there's the chemical cascade, neurotransmitters gone so haywire that they are released in quantities and mixtures that not only can't support consciousness but are downright toxic.

Doctors will try all sorts of desperate things to save your brain. They may cut holes in your skull to suck out the blood or remove entire sections of bone to give the engorged tissue some room to expand. They may give you drugs or chill your body in an ice bath or put you in a hyperbaric chamber to slow down the destruction wrought by your own metabolism. They'll do all of this knowing that none of these measures is necessarily better than any other at restoring consciousness, that it is nearly impossible to predict accurately what your ultimate fate will be or how long it will take to find out. So most of what they do is wait to see what happens.

Members of your family are waiting, too, and they are faced with decisions no one should ever have to make. A brain injury severe enough to put you in a coma for more than a few hours is virtually guaranteed to force a discussion about whether to keep you alive.

Your family will depend on your living will if you have one, on what you might have told them about this possibility, and on their own sense of what makes life worth living. But they will also be dependent on doctors to give them some idea of your prognosis. The doctors, in turn, will rely on the diagnoses that constitute the *severe disorders of consciousness.* In addition to brain death, these include the *vegetative* state—in which a patient's eyes are open but there are otherwise no signs of consciousness (if this lasts for a year, he or she is said to be *persistently vegetative*)—and the *minimally conscious* state, in which a patient has some awareness and a limited ability to respond to commands, maybe with an eye blink or a squeeze of the hand.

Whatever they tell you, however, probably won't include the fact that aside from vague generalities about the importance of the connections among neurons, doctors don't really know what makes us conscious; where exactly the switch is that turns us from animate, metabolizing objects to human beings with awareness, the subjects of our own lives; or how all those neuronal connections add up to the experience of being an "I." For the time being anyway, those questions remain the province of philosophers.

That doesn't stop doctors from presenting their diagnostic distinctions as science-based and rock solid, however, or from claiming that their training qualifies them to diagnose a life as no longer worth living. Candice Ivey's doctor claims not to remember being angry at her mother (or much else about the case, for that matter), but it's easy to understand his anger. By the time you get to be a neurosurgeon in a teaching hospital, you've very likely forgotten the tenuous nature of these diagnoses, the way they gloss over murkier distinctions. Which is perhaps as it should be: a doctor paralyzed by unresolvable moral dilemmas might just as well hang up his or her stethoscope.

But if being certain where others hesitate doesn't do much for bedside manner, it would be a mistake to accuse doctors like Candice's of failing to revere human life, let alone of having a death wish. What they do have is a conviction that consciousness is what makes life worth living and that the brain is what gives us consciousness. This isn't a terribly radical notion, not when every week brings a new finding about the neural basis of behavior and experience. Indeed, it's only when someone like Alan Shewmon points out that if we take these convictions as the start of a syllogism and follow them to their conclusion—that without a functioning brain, there is no human life—then it is disturbing, to say the least, that the propositions are speculative, that we have no way of *knowing* that a vegetative state is the absence of experience, or, for that matter, that it is not pure bliss.

This false but consoling certainty is at the heart of our noble lies, just as Plato intended it would be for the citizens of the Republic. And it is not an accident that these fictions are grounded in the material world, for we have come to think of that which we can grasp with our senses as our best source of certainty. This is why science attempts to reduce complex phenomena to their material components in order to generate knowledge. But, as molecular biologist Carl Woese points out, this can lead us in at least two different directions:

> Empirical reductionism is in essence methodological; it is simply a mode of analysis, the dissection of a biological entity or system into its constituent parts in order to better understand it. Empirical reductionism makes no assumptions about the fundamental nature . . . of living things. Fundamentalist reductionism . . . on the other hand is in essence metaphysical. It is ipso facto a statement about the nature of the world: living systems (like all else) can be completely understood in terms of the properties of their constituent parts.

The idea that consciousness is only what the brain does is a prime example of fundamentalist reductionism. By the time this metaphysics gets to the ICU, the fact that it is speculative has long ago been forgotten. It has become a brute fact, delivered in the most horrific circumstances by a believer so true that he doesn't even know he's espousing belief. Which may be exactly what you would want under the circumstances: certainty as a shield against the raw presence of life.

It is also no accident that the people most likely to take a stand against fundamentalist reductionism are fundamentalists of a different stripe—the Alan Shewmons of the world, whose own metaphysical commitments predispose them to seeing the metaphysics lurking in scientific certainties. But you don't need theology to grasp the fictional nature of noble lies such as the diagnosis of persistent vegetative state. All it takes are the occasional diagnostic errors or the mysterious awakenings or the itinerant doctors with their electrified cuffs who make it clear that when it comes to these diagnoses, and the prognoses that follow them, doctors aren't as sure as we all might wish they were.

ED COOPER LIKES TO SHOW OFF Kinston, North Carolina, where he has lived and practiced orthopedic surgery for more than thirty years. As he drives me around town, he doesn't seem to notice that it's a down-at-the-heels place with a dying Main Street whose most thriving businesses appear to be the bail bondsmen near the courthouse. "When the truth won't set you free, we will," one sign proclaims. The decay seems irreversible, but then again, Cooper is not one to give up on hopeless cases.

He credits his wife, Mary-Bryan, with discovering the effects of electrical stimulation on impaired brains. In the mid-1980s, he was using a neurostimulator to relieve spasticity in the limbs of microcephalics, whose small skulls cause mental retardation and poor muscle control. Mary-Bryan, who once studied to be a neurosurgeon, was watching a videotape of a stimulator session.

"I guess I saw it as a mother would," she told me. "The boy was looking around his room instead of staring blankly at the wall, suddenly smiling when people walked in the room."

Cooper had already observed that when he placed the cuff on one arm, the opposite arm also responded. He concluded that the electricity was making its way into the brain, crossing into the opposite hemisphere, and stimulating the central nervous system on its way. He began to wonder about the effect this might have on unconscious people. "I thought, if someone were normal and able-bodied but were in a coma, maybe this would make a difference, maybe help wake them up," he said. "It was like maybe we could reboot the brain."

Cooper had just gotten approval to research this hypothesis at East Carolina University (ECU) and the University of Virginia when a fellow mourner at a wake told him about a girl in a coma at ECU's Pitt Medical Center. "I got right out of that line," he said, "and went to find her."

Cooper, who is sixty-three, didn't charge the Iveys for his services, nor has he charged any of the more than sixty patients on whom he's tried right median nerve stimulation. He's asked for nothing for the late-night phone calls or the testy consultations with suspicious doctors, or even, in some cases, for the stimulators themselves, which cost $1,400, and of which he's donated a dozen to ECU. He's seen people improve after as much as ten months in a vegetative state and has managed to gather official data on thirty-eight subjects enrolled in his ECU study and another study at the University of Virginia. While no one has recovered as spectacularly as Candice Ivey, his published data say that people emerge from comas more quickly and then get better at a rate far beyond what would normally be expected, leaving the hospital under their own power, with less severe disabilities, at twice the rate that would otherwise be expected from the nature and extent of their injuries.

Still, Cooper knows that thirty-eight patients is a tiny sample, especially in a field where so little is understood and in which unexplained spontaneous awakenings, even after long periods of unconsciousness, are not uncommon. But despite being published in journals such as *Brain Injury* and *Neuropsychological Rehabilitation*—and despite the fact that electrical stimulation has been used with success on unconscious animals and with Parkinson's disease patients and has been shown to stimulate the arousal centers in the mammalian brain—his work has yet to attract the attention of mainstream researchers, whose closer ties to universities or industry might lead to a bigger study.

So in the meantime, Cooper, who is semiretired after a bout of cancer last year, does what he can to spread the word. "On Google, I have an alert for 'brain stem injury' and 'teenage coma,'" he told me. He calls hospitals and faxes doctors; he has even tried to intervene in high-profile cases like those of Henri Paul, Princess Diana's chauffeur, and Marie Trintignant, the French actress who died after being beaten into a coma by her boyfriend in 2003. The patients and doctors rarely respond, however, and Cooper remains a small-town doctor, stuck on the margins of medicine, frustrated and sometimes bewildered. "It's like I'm George Washington Carver," he said, "thinking that I've discovered something that no one wants to hear about. It's so easy. Why don't people just use it?"

COOPER MAY BE WITHOUT HONOR in his own home, but mention his—or Candice Ivey's—name at the Fujita Health University Hospital, just outside the industrial city of Nagoya in central Japan, and the local neurosurgeons light up with recognition. He's been to Fujita a few times, collaborated with the Japanese doctors on a book chapter, and presented at their conferences. (He's also collaborated with doctors elsewhere in Japan, as well as in Taiwan and Shanghai, a city whose sudden upsurge in automobile ownership has led to a huge increase in severe head injuries.) The Japanese are glad to have a fellow traveler in the United States, but they're quick to point out—politely, of course—that they've been doing this work for more than twenty years now and have treated hundreds of patients.

The Japanese also use a more spectacular method: they implant the electrodes right into the spine. That's what Isao Morita is doing today. Trained at the Cleveland Clinic, he's a neurosurgeon who wears his hair in a brush cut and speaks passable English. The patient, Katsutomo Miura, lies facedown on the table. He's anesthetized, but it's not clear to me why, since he was already unconscious when he was handed through the airlock doors separating the sterile surgical wing from the rest of the hospital. In fact, he's been unconscious for nearly eight years. He was twenty-three when an ambulance crew found him bleeding and unresponsive in the road near his home in Osaka, next to his wrecked motorbike and his helmet. His legs were shattered, and one of them is now permanently bent at the knee, as if he was frozen in place while he was about to run away. It sticks up from the table, making a little pup tent under the blue surgical drapes.

"Yoroshiku onegai shimasu" ("Thank you in advance for your cooperation"), Morita says, and waits for the five-person surgical team to respond in kind before he slices into Miura's neck. It takes twenty minutes of cutting and cauterizing, of spreading muscle and clearing away gristle and blood for Morita to burrow down to Miura's spine.

"C-5," he announces to me, a little triumphantly, as he points into the cavity he has created, a wound that you would expect to see only on a battlefield. Peering over his shoulder, I can see the vertebra that was his target. It is pure white and glistening. Morita takes a pneumatic drill and tunnels along the spine, up toward Miura's head, explaining that so far this is exactly how a disk surgery would go. I resolve to take better care of my back.

Morita tries to push an inch-and-a-half-long, quarter-inch-wide, flat metal bar into the tunnel, but it won't go. He drills and pushes four more times until the electrode finally slips into place along the second and third cervical vertebrae. He snakes a wire under Miura's skin to another incision he's made between Miura's shoulder blades. Meanwhile, another doctor has been working at Miura's waist to create an internal pouch for the battery pack that will power the electrode on his spine. Now she runs a wire up to the opening Morita has made, and, using four tiny screws, Morita splices the two wires to complete the circuit. Once the swelling from surgery goes down and they switch the implant on, it will send a steady trickle of electrical pulses through his spinal column and into his brain. The hard part over, the surgeons begin to chat easily as they sew the equipment into place, even laughing a little bit at the anesthesiologist, who has dozed off at his station.

I've already seen this kind of operation on video. It was part of the private PowerPoint presentation I got the day before the surgery from Tetsuo Kanno, Morita's mentor and the originator of the surgery. He told me about how he, like Ed Cooper, had discovered the virtues of the dorsal column implant accidentally—in Kanno's case, when he was using it to stimulate muscles in stroke patients. He outlined his theory about why electrical stimulation works: patients receiving stimulation have higher levels of dopamine and norepinephrine activity and increased blood flow, all conditions associated with arousal. He showed me the statistics on the 149 people he and his staff have treated, citing one study of patients who had been unconscious for an average of nineteen months. Even though a vegetative state is presumed to be permanent after one year, 42 percent of Kanno's patients showed significant improvement. It's never too late to try stimulation, he said, insisting that even a guy like Miura stands a chance. He'll have to come back every year or so to get his batteries replaced—in fact, there are a couple of patients in the hospital right now for just this reason—but it's likely that if the electrical current keeps flowing into his brain for long enough, maybe years, Katsutomo Miura will make "some recovery."

Which is either good news or bad news, depending on how you feel about Kanno's definition of recovery. Most of the implant recipients, he thinks, achieve a minimally conscious state, able to muster small but unmistakable signs of awareness. "Maybe the patient just smiles or follows with their eyes," Kanno said. Doctors at other hospitals in Japan have reported similar results using deep brain stimulation: patients who improve to the point that they are severely disabled rather than entirely unresponsive.

But this is enough for Mariko Miura, who spent $30,000 on her son's implant surgery, to hope for. The day after the surgery, she told me, through a translator, that her son was calm and comfortable. She is sure that "if he could just show what he feels, yes or no, maybe blinking once or twice, maybe holding hands, maybe a smile, that would be great." The doctors say that giving Mariko

Miura this hope is the reason they are doing this surgery even though Katsutomo's MRI shows that his right cerebral hemisphere is almost entirely atrophied.

"There is no medical indication in this case," Morita said. "This surgery is socially indicated. It is the family's decision whether they want to go on, and our job to do what they wish."

These doctors know how strange this kind of reasoning, and how meager the expected gain, sounds to an American ear. But then again, in the United States, these patients tend to get warehoused in nursing homes, while in Japan, where nursing homes are few, they live with their families; Mariko Miura tends to her son almost single-handedly, clearing his breathing tube every half hour round the clock. "U.S. doctors say that it doesn't mean anything," Kanno told me. "But even if the patients can't talk, if they just look out when the family comes in the room, it makes the family very happy." But, then again, he said, "You are very dry people in America, dry and cool. Here we are very wet and warm. You see just a body; you say: okay, stop feeding it. But we think a person in a vegetative state has a soul."

BACK IN THE UNITED STATES, bioethicist Robert Veatch believes that the problem of what to do with patients in persistent vegetative states (PVSs) can be solved without worrying about the state of their souls. For that matter, he believes that the issue can be settled without family members agonizing over what treatments to allow or how best to implement a living will. Veatch, a professor of medical ethics at Georgetown University's Kennedy Institute of Ethics, thinks that people who have permanently lost consciousness should be declared dead—at least, if that is what they have said they want. Veatch has proposed that states add "conscience clauses" to their statutes governing the declaration of death, which would allow us to make our own decisions about when we should be considered dead—within limits, anyway.

"If you were up and walking around and said you wanted to be treated as a dead person, society can't really handle that," he told me from his office. "And if your body was putrefying, there is a public health interest in treating you as dead even if you had insisted that you would be alive in that condition." But a diagnosis of PVS would be one of the conditions, which would also include brain death and cardiorespiratory arrest, that people could choose as the criterion of their own deaths. (Some states, such as New Jersey, already have opt-out laws that allow people to specify on religious grounds that they are not to be declared brain-dead.)

Veatch points out that the primary importance of the death diagnosis is that it dictates how we will treat someone. Giving people a choice about how they want to be treated, and at what point, is an acknowledgment of the fact that when life ends, not unlike when it begins, is an open question, and not one that science is going to solve. "In philosophical disputes, we give people the right to make their own choices and their own commitments. You ought to be able to pick when you want to be treated as a dead person."

If Veatch's proposal were law, then PVS patients who had chosen that condition as a criterion of their own deaths would be permanently unconscious not because of the nature of their injuries but because of how they are treated. And some doctors think that even without a conscience clause, death in cases of devastating brain injuries is often iatrogenic in just this fashion, that a PVS diagnosis is a self-fulfilling prophecy. This is particularly troubling, according to Joseph Giacino, a rehabilitation psychologist at the New Jersey JFK Johnson Rehabilitation Institute, because in more than 40 percent of cases, doctors render a diagnosis of PVS when the patient is minimally conscious or even less impaired than that.

Devastating brain injuries, Giacino thinks, often bring on a "rush to judgment": doctors tell families that the outlook for their loved one is hopeless, and families decide to discontinue treatment before the diagnostic picture is clear, before the patient can rally, or before intensive rehabilitation can rouse him or her. The patient in turn becomes part of the grim statistics that doctors cite in making prognoses, which leads researchers to assume that there is no point in investigating possible treatments. This occurs despite the occasional headline-grabbing awakening, and despite scientific evidence that, at least in the minimally conscious, people who appear to be unresponsive continue to show measurable brain activity and even regrowth of the connections among neurons. (Giacino has led the effort to establish the minimally conscious state as a diagnostic category.)

Giacino's colleague, Joseph Fins, an ethicist at Weill Cornell Medical Center, calls this vicious circle "therapeutic nihilism" and argues that it is an unintended consequence of the 1976 New Jersey Supreme Court ruling that Karen Ann Quinlan, a twenty-two-year-old who had suffered a severe brain injury, was beyond sentience and thus could be allowed to die of starvation. "The right to die was established on the grounds that we weren't giving up anything by allowing people like this to die," he told me. "We've spent a long time allowing these people to die, but maybe they are more interesting than we thought. Maybe they deserve more intellectual, diagnostic, and therapeutic engagement than we have acknowledged."

Giacino and Fins are part of a team that is trying to engage these patients by investigating electrical stimulation as a treatment for devastating brain injuries. They have tested various hypotheses on animals and laid out the ethical case for experimenting on people who can't give consent, and in 2007 they announced the results of an experiment conducted on a thirty-eight-year-old man who had been in a minimally conscious state for more than six years following an assault. The doctors used deep brain stimulation, sending electricity directly into electrodes implanted in individual neurons, and they reported that after six months of treatment, the patient was able to name objects, to control his swallowing reflex well enough to take his food by mouth instead of by feeding tube, and even occasionally to utter five- or six-word sentences. These results, the doctors wrote, "should motivate research to elucidate the mechanisms of recovery and to facilitate the identification of patients who might benefit" from the procedure.

The American doctors rarely acknowledge the work of Kanno or the other Japanese surgeons. In fact, Giacino calls their research "bad science" and is openly skeptical of their claim that patients who were vegetative for years before treatment were roused. He thinks that at least in some cases the Japanese doctors have misdiagnosed their patients and that they were minimally conscious, not vegetative, in the first place. But in Japan, such fine distinctions are unnecessary because there is no right to die; Kanno told me that when families become exhausted or overwhelmed with their duties, physicians advise them to leave a window open and wait for the patient to contract pneumonia. For Giacino and Fins, on the other hand, the differential diagnosis is crucial, not only because it guides who is chosen for treatment but also because they want to drive home the point that some patients are indeed beyond help. They both cited Terri Schiavo as a prime example and said that, at least in those cases, they do support what Fins calls "the hard-won right to forgo life-sustaining therapy." If this seems like protesting too much, it is only because they recognize that their work has the unsettling potential to undo the medical and social consensus that emerged after the Quinlan case.

"We're asking to engage the very people in whom the right to be left alone was first established," Fins told me—a move bound to discomfit "proponents of the right to die." And these patients may respond with only minimal improvement—just enough to show that their conditions are not permanent, but not enough to overcome objections that they are still too impaired to lead meaningful lives. "The question is always how good is good enough? It's going to make these decisions much more complicated," Fins said.

Things will get particularly complicated if evidence shows, as both Cooper and Kanno believe it will, that electrical stimulation pushes people out of persistent vegetative states and into minimally conscious states. It will raise uncomfortable (and perhaps unanswerable) questions about whether the patients are actually better off with a small amount of consciousness restored. We can't ask them," said Kanno; whereas Giacino at first claimed, "They don't get aroused enough for self-awareness" and then admitted that he had no way of knowing that. But perhaps more important, if research on electrical stimulation shows that PVS is not entirely hopeless and irreversible, then the diagnosis that has functioned as an important rationale for ending life support will no longer provide moral clarity. "If it turns out that we have a critical mass of evidence," said Giacino, "people are going to have to really think about what this all means before nonchalantly pulling the plug."

OF COURSE, IT IS HARD to imagine that anyone makes that shattering decision nonchalantly. But perhaps people do take as certain some things that are not quite true—namely, that people in vegetative states cannot be treated. This, of course, was the pivot on which the Terri Schiavo spectacle turned: that her diagnosis wasn't as final as it seemed, that maybe that little smile meant something, and that starving her might be murder rather than mercy. As it happens, she would have been unlikely to respond to any form of electrical stimulation; cases in which the brain has been deprived of oxygen, rather than injured by force, are the hardest to treat. But accident victims fill emergency rooms, and it is hard to picture how much more tortuous our decisions will get if new truths about electrical stimulation displace old certainties about hopelessness.

Even with the current guideposts, the complexities seem mind-bending. Just ask Candice and Elaine Ivey. With Candice's impaired short-term memory, her lack of stamina, and her difficulties keeping friends, her life—one of the best possible outcomes after so severe an injury—is still immeasurably harder than it was. "God's allowed me to do a lot of good things," she told me. "But I remember what life used to be like and what I used to do mentally and physically, and I wouldn't want to do this again. If this ever happens again, I want them to terminate me."

When I asked Elaine Ivey about this, she drew deeply on her cigarette before she told me that Candice has often been depressed and suicidal since her injury. "It goes through my head every day. If I had let her die, she'd at least be at peace. And I keep thinking there has to be a reason for this, her life will turn around, but when it doesn't happen . . . I mean it's been twelve years now."

Things are no simpler in Katsutomo Miura's hospital room the day after his surgery. He's entirely still except for his lips, which are rooting ceaselessly like a hungry infant's. His mother, who is bustling over him, leans into his face, squeezes his cheek, and talks to him. I realize that she is introducing him to me. "My son and I, we are one person," she told me earlier, and as if to prove the point, she picks up his right hand and extends it for me to shake. It is warm and wet.

Not for the first time in my three days at Fujita, I'm reminded of another doctor who much more famously applied electricity to a lifeless body to animate it. Of course, Victor Frankenstein's wish to cheat mortality is the impetus behind all medicine, but you don't often see its monstrous implications as clearly on display as it is in this poor man suspended by good intentions between two worlds.

"We produce these patients," said Kanno, by way of explaining his commitment to his electrical stimulation project. "It is the dark side of neurosurgery."

Unintended consequences, and the impossibility of unraveling them, are on my mind as I finish my visits with implant patients and their mothers. No one seems to be much concerned about what this is like for the patients, and I'm wondering why these women can't see that their children are gone forever, why they can't, as we say here in the United States, move on. I want to say something like this to my translator as we get into the elevator, but there are tears in her eyes. "They're so well loved," she says, and I can't help but think that I am not only on the other side of the world but also on the other side of our beliefs about what makes a life worth living, that I am grasping the moral chaos that will ensue if science proves these doctors right.

7

MORTALITY: WE'LL ALL WAKE UP TOGETHER

Y OU DON'T HAVE TO SPEND too much time with a geoduck clam to understand why the species has a reputation for being an aphrodisiac. The geoduck's (pronounced "gooey-duck") siphon, what clam eaters everywhere call the neck, is a tube-shaped affair protruding from a bifurcated shell, eight or twelve or sixteen inches long and an inch and a half thick, and it looks for all the world like a huge uncircumcised penis protruding from an equally outsize scrotum—or, as *Dirty Jobs*'s Mike Rowe told Jon Stewart one night, like a phallus attached to a shoe. The siphon shrinks away when you grab it, but you can't take it personally. The clam probably thinks you're an otter or some other predator trying to dislodge it from its burrow in the seabed. If it escapes this fate, and if it is lucky enough to avoid the dogfish that detach it from below or the starfish that like to nibble on its shaft, the geoduck will stay there sucking down plankton, excreting filtered water, and, if it's a female, laying as many as a billion eggs, for a century and a half or so, at which point it might weigh fifteen or twenty pounds and be worth $500 or $600 at retail.

Most geoducks end up in Asian markets and restaurants, but they grow almost entirely in the deep, cold waters off the North American Pacific Coast. Marine biologists estimate that the adult population of the Puget Sound alone is at least 300 or 400 million. And the sound offers something else: a unique set of laws that allows private ownership of the nutrient-rich and easily accessible tideland, an environment that is perfect for aquaculture. So a guy like Jim Gibbons, a tall, silver-tongued fifty-year-old with a thick salt-and-pepper mustache, can take a clam seed or two million, plant each of them in its own four-inch plastic pipe (which protects it from the starfish and the otters), and after five years reap a bumper crop of well-endowed clams ready for the sashimi knife. Seattle Shellfish, Gibbons's company, has been doing exactly that in the southern Puget Sound for ten years, and in 2006, his crew of fourteen harvested 350,000 pounds, turning a million-dollar profit for Gibbons and his investors.

He may be one of the biggest producers of geoducks in the world, so successful at it that his physician wife recently retired from her medical practice and teaching post. But when he's standing at the wheel of his seventeen-foot runabout and giving you the grand tour of the geoduck business—the divers hauling bushels of his clams to the surface off the shore of one of the archipelago's islets, the beds a few miles south where siphons stick up out of the mud like so many inverted carrots, the beach that some people think should be public rather than leased to the clam farmers, the enclave of multimillion-dollar houses (one of which he plans to move into in the next few weeks), the site of the most recent rockslide—he doesn't seem so much a successful entrepreneur looking after his business interests as a hooky-playing kid thrilled to be cruising the sound on a Monday afternoon, even a gray and misty Monday like this one. Not that Gibbons has ever been much for the workaday world. His last project, also a success, was rebuilding a run-down hotel northeast of Seattle and turning it into a yuppie getaway. When he did that, he knew as little about building rehab and the hospitality industry as he once did about aquaculture and life on the water.

Well, maybe Gibbons is still learning about boating. He's showing me his brother-in-law's hot tub, up a sloping lawn from the shore. It's a beauty all right, but before we can imagine what it's like to end a winter night with a warm soak, there's a loud bang and the boat shudders to a stop.

"I guess we hit a rock," Gibbons says. He seems calm, and we are only a few hundred feet from shore with a paddle tucked into the gunwale, so I decide to stay calm, too, even when he pulls up the motor to reveal a propeller with one of its blades folded over. He drops the motor back into the water and turns the key. The engine catches and runs—not quite evenly, a little surge in the middle of every revolution.

"Uh, Jim, you might not want to run that motor with a bent prop," I say. I'm remembering a time when I was a boy and nearly died, or so it seemed, when a friend's father ran his Chris Craft over a log. The boat heeled over so far, it almost dumped us into the ocean; then the father ran the engine until it seized. But that boat had two motors, and all we have is our 70-horse Yamaha, and suddenly Puget Sound doesn't seem quite so cozy. "Run that shaft out of balance long enough, and . . ."

"Aw, it'll be fine," he says. And opens his coat to the drizzle, stands up and sticks his head into the wind, and wheels us back into motion, our excursion hardly interrupted. And you see it now better than you have all day—better than when you heard all of his stories about how nothing stands between him and what he wants for very long; better than when he was speeding down the winding, rain-slicked roads of the Olympic Peninsula, dodging logging trucks in his 1990 Mitsubishi, its exhaust system barely held in place, its old tires struggling to hold on in the curves; better even than when he backed his boat down the ramp on a trailer hitched to an eighties vintage pickup with a kaput alternator, which he knew wasn't going to start again if it stalled out with its ass end submerged in the rising tide, and through it all remained calm and relaxed and confident. You see that it's all in the launch, in throwing himself toward the goal, finding his limits, and then going past them.

Which, I'm just figuring out, is why I've come to see him. Because what other kind of person kicks down $80,000 to go past the biggest limit of them all? That's how much Gibbons has arranged to give to the Alcor Life Extension Foundation in Phoenix. In return for that, Alcor's doctors and technicians, the *readiness team,* will, when Gibbons *deanimates*—or, as the rest of us say, when he dies—put him on ice and under the Thumper, a mechanical CPR device that will keep some oxygen flowing through his body, forestalling ischemic damage until he can be gotten to Alcor's surgeons, where they will decapitate him, pump his brain full of a high-tech antifreeze, cool him to below freezing, submerge his head in a shiny stainless-steel tank, and maintain it in a −170-degree Celsius bath of liquid nitrogen until he can be thawed, revived, and cured of whatever it was that made him ready for the team in the first place. (For $150,000, Alcor will skip the decapitation and place your whole body in suspension.) Should readiness occur in unforeseen circumstances, Gibbons is wearing a necklace and a bracelet, like Medic-Alerts, with directions on what to do and whom to call until the team can get there.

"Because I love life," he keeps saying to me when I ask why, as if the answer is self-evident to the question I've flown out to Olympia to ask, as if, in fact, it is a stupid question; and now I see what it is that he loves—sheer experience, of which he can never have enough. The fact that mortality means that there will be an end to experience, to trying new things—it is an affront to a man like Jim Gibbons. And even if the chances are slim that he will die the right way and get the postmortem attention he needs, that his organs will withstand the infusions and his long immersion in the tank, that future doctors will know how to revive and heal him, or that Alcor itself will survive that long, even if this convergence of factors is as unlikely as threading a camel through the eye of a needle, he insists that it's worth the shot.

"At least I'm not in the control group," he says and starts to tell me about the six screenplays he's written.

I'm worrying that he's not so much evasive as just plain unreflective—not a good thing for someone with a story to write. But in any event, we've got something new and more pressing to talk about. It's been an hour or so since our encounter with the rock. We've been tracing a lazy circle around the ragged coastline, as small storms welter and pass overhead, turn the mist to rain, and whip up the sound before they move on, dragging enough clouds in their wake to reveal the sun for a moment. We're just about to round the horn of Harstine Island and make the ten-mile run back to the marina when the motor, whose surging and sputtering have been on the increase, gives one last cough and dies.

Gibbons cranks the motor, which turns over but doesn't catch. He cranks some more, and now it is totally silent. He picks up the engine cover, jiggles some wires.

"I'm a mechanical idiot," he says, a hell of a time for such a confession, but I don't think even a mechanical genius could help us right now, not unless he has a spare motor with him. The prop shaft is busted, the motor is as deanimated as a doornail, we're ready-for-the-team in the water about two hundred yards off Harstine's northern tip. On the little beach is an orange windsock, stiff in the breeze, and a picnic table, with a house, no doubt, not too far back through the woods. But we're drifting away from this patch of civilization, farther into the deep, silver-green water.

"Maybe we should row to that beach," I say.

"Oh, that oar." Gibbons shakes his head. "That's a whole other sad story."

I pull it out of its slot in the gunwale and come up with nothing but handle. The paddle has broken off. I toss it to the deck. I don't want to hear that story.

Gibbons is standing up with his phone to his ear. "This is where it comes in handy to be the president of the company," he says while he waits for his connection. He's talking to his office manager. About the harvest, about the tides, and, finally, just before I'm going to scream at him, about our situation. "You're gonna have to come and rescue me," he says, and he's somewhere between sheepish and resigned, as if this is not the first time. "We're just off Harstine, couple miles up from the bridge." They make arrangements. Gibbons signs off and settles back into his seat. He gives the key one more cursory twist.

"So how long is that going to take?" I ask.

"Let's see. They have to put the boat on the trailer, take it to the ramp, come up here—I don't know, maybe forty-five minutes or an hour."

We're drifting northwest now. It's only a couple of miles west to the other side of the strait between Harstine and Stretch Island, but the current is running hard from south to north. I can't tell where it will sweep us to, but it sure looks like open water. And running from the south with the tide is another mini-storm, the gray skies turning bruise blue and streaked with rain, whitecaps rising underneath. The chop is increasing ahead of the squall, a foot, maybe eighteen inches—nothing to pay any mind to, unless you're on a rudderless ski boat with only a foot of freeboard at the transom, in which case you are up the sound without a paddle. The *Gilligan's Island* theme song runs through my head.

"You got life jackets on this thing, Jim?"

He points to a spot in the tiny cuddy cabin. He doesn't move toward them but picks up his legs so I can get by. Earlier, Gibbons had asked me what I wanted to do with him today. "Just think of me as a hitchhiker," I said. "Just along for the ride while you live your life." I'd do whatever he did, I said, which is how I ended up visiting the hatchery that nursed a zillion baby clams (many of which were infected with some awful bacteria, threatening a crop five years in the future), stopping in on his wife and kids to pick up hip waders and slickers, helping him jump-start the pickup, and even when I wanted to tell him to drive slower, to be a little more attentive to the family, to take the time to charge the truck battery, I had stuck by my hitchhiker's creed. But now I can't stand another minute of whither thou goest. I reach into the locker and grab an orange vest. I hesitate, then take out another and hand it to him. He puts it off to his side while I strap on mine. It's more symbolic than anything else. There's no way I'll survive in the 45-degree water long enough to swim to shore. But I don't want my wife and son to think I didn't try, that I was heedless of their possible grief.

"If we have to go in," Gibbons says, "make sure you take off those hip waders before they fill up."

"If we have to go in, Jim, I'm taking that medallion off from around your neck and putting it on mine."

Which, I admit, isn't the nicest thing to say. And I'm not sure I'm kidding. It's a little surprising to hear this coming out of my own mouth, but on the other hand, I've been doing the math. I'm in a possibly life-threatening situation with just the wrong kind of person—not only a risk-taker, a self-confessed idiot with a limited patience for details and maybe a tenuous hold on his impulses, but a guy who thinks he is going to live forever, or, to be more precise, who thinks he is going to go into hibernation until medical science catches up, as he is sure it will someday, with his desire to be immortal. He's sitting there sprawled in his vinyl seat, awaiting one form of rescue or the other as if he has all the time in the world. Which, of course, he thinks he does. Not only that, but maybe the only thing better than having the readiness team right there with their dry ice and their Thumper is to deanimate in frigid, ischemia-unfriendly water, where, as long as he doesn't get eaten, he'll be all set to wake up in a better tomorrow.

Myself, I'll settle for a hot shower and a warm bed tonight, and I am surprised by my snarl. Because I came to Washington with the idea that mortality was just one of those things that you have to accept, that three score and ten or maybe four score at the outside was about all you can ask for, that all this grasping for more years was just so much vanity and terror, and that if there's anything left in us of the natural world, it's our limited life span and the necessity of returning to dust, a tonic of humility against the arrogance of modern life. But out in the middle of Puget Sound, I'm thinking that these Alcor folks might be onto something. And it's not just a foxhole conversion, my fear and loathing of death inclining me toward their side. It's also something I've learned since I got involved with the immortalists: that nature may not be as bound and determined to kill us as we have come to think, or as it seems at this moment.

OUT OF THE BLUE the other day (don't these things always happen out of the blue?), my nearly ten-year-old son said, "I wish I could be dead for just a minute and then come back."

Me: "Oh, yeah? Why's that?"

Him: "Because then I could see what it was like."

Me: "And why would you want to do that?"

Him (exasperated at my denseness): "So that I could know if it's better or worse than being alive."

Rather than push the panic button—*Omigod, my kid is weighing the costs and benefits of suicide*—I decided to take this as his way of fashioning an experiment that perhaps all of us would like to conduct. Our fear of mortality is more than a fear of the unknown. At least for those of us who don't believe in an afterlife into which we can bring our memories and experiences, it comes from a dreadful certainty that there is an answer to my son's question: that oblivion awaits us. The mind reels at the prospect of its own disappearance into an infinity of nothingness, but until someone invents a real-life version of Dylar, the fictional drug that in Don DeLillo's *White Noise* removes the fear of death, we are just stuck with the terror, left to scream like the anguished man in Edvard Munch's famous painting.

Unless, that is, we don't have to die. What if it turns out that the universal diagnosis—life as a terminal condition—is one of our noble lies? Certainly it has the hallmark features: it helps us get on with making an orderly life by imposing a narrative frame on our lives, a sense of beginning, middle, and end, and the necessity to strive to build some kind of trajectory through time. It makes it possible to accept a condition that seems unacceptable—our own deaths. And it papers over an important moral question—what are the limits of medicine, and, more generally, what are the limits of our ability to change the world? We live in the land of self-improvement, but the very idea of self-improvement implies that how nature made us is not enough. This impulse gives us polio vaccines, antibiotics, and teeth that last our entire lives. But it is at its heart the impulse to reject nature's terms, to hack the source code, and, ultimately perhaps, to live forever. The conviction that we must be mortal, that medicine cannot go beyond this limit, applies a counterweight to that impulse, helps to prevent it (to the extent that it has been prevented) from becoming pure hubris.

It is for just this reason that the current President's Council on Bioethics has warned against pursuing immortality. Mortality, the council has pronounced, is important precisely because it imposes limits on our attempts to improve on nature:

> A flourishing human life is not a life lived with an ageless body or an untroubled soul, but rather a life lived in rhythmed time, mindful of time's limits, appreciative of each season and filled first of all with those intimate human relations that are ours only because we are born, age, replace ourselves, decline, and die—and know it. It is a life of aspiration, made possible by and born of experienced lack, of the disproportion between the transcendent longings of the soul and the limited capacities of our bodies and minds.

Lose mortality, the president's men warn, and we may lose all aspiration.

The rhetorical contrast between this presidential committee and the one that worked out the details of brain death couldn't be more stark: Bush's council has surged into the philosophical country where Carter's commission feared to tread. The leader of this charge is council chairman and ex-doctor Leon Kass, a philosopher and a bioethicist who has written extensively about the way the modern world casts us adrift from any ethical moorings, leaving us to pursue our "open-ended desires and ambitions." This amounts to a thoroughgoing critique of more than just medical technologies. It is a broadside against modernity, or at least against the modern tradition of liberal democracy and its declaration that freedom and equality are the ultimate purpose of human life. But, Kass argued, we retain the vestiges of moral sensibility, and they are to be found in the revulsion inspired by such notions as living forever, a reaction Kass calls the "wisdom of repugnance." Were we to heed that wisdom, Kass said, we would reject the practices that our noble lies currently justify: brain death, the reliance on psychiatric distinctions to replace moral distinctions, the use of drugs to improve our moods, the "nonchalant" removal of life support, and homosexuality, not to mention steroids, embryonic stem-cell research, and contraception. (Kass's positions on these matters have led some technology enthusiasts to call him a "bio-conservative.")

Not all repugnance is created equal. After all, some people are repelled by such ideas as interracial marriage, homosexuality, and "taking your organs to heaven." But Kass thinks there is a place we can turn to guide our disgust: the Bible and, in particular, the Book of Genesis, which provides "insights mysteriously received from sources not under strict human command." Reason, in other words, is not sufficient to establish the premises from which we can figure out our limits; instead, perhaps, "we should pay attention to the plan God adopted as an alternative to Babel, walking with Father Abraham." Noble lies, in this view, are all that our rationality can provide to ground our morals. But there is an alternative to these contingent and changeable verities: revelation. And when it comes to immortality, the Bible is clear: it is the province of God (and, in the New Testament, His son).

Kass's solution doesn't work if you're not a believer, and at least one critic has suggested that "the truths he traces to revelation are foregone conclusions based on his own philosophical stance toward life." Such is the fate of preachers in an age that has rejected universal faith. But there may be a deeper problem with Kass's attempt to find limits in revelation. His intellectual mentor (and the philosophical godfather of the neoconservative movement) was Leo Strauss, and while Strauss thought the Bible was a Good Book, he had a peculiar stance toward religion, one that should be familiar to you by now. He thought that it was a noble lie (or, as he wrote, a "pious fraud"), a stabilizing structure that the masses needed in order to prevent them from falling into nihilism. (In Strauss's world, the wise were those who could accept the "natural hierarchy" without needing to see it as the work of God.)

So we may not be able to choose truth but only to decide which fictions we want to live by. Or to put this another way, when it comes to medical technologies, there may be no other limits than those imposed by contingencies, particularly by pragmatic contingencies—the expense of technologies, their technical complexity, their usefulness, and their appeal. And, at least if you're not looking at it from the perspective of natural selection, dying is really impractical. Which is why it is fitting that one of the early founders of the republic was also one of the first Americans to seriously consider the possibility of living forever.

ON A TRIP TO LONDON in the early 1770s, Benjamin Franklin presented his host with a bottle of Madeira. When he decanted it, three dead flies tumbled into the glass. They had, Franklin figured, flown into the Madeira when it was bottled back in Virginia. If he was embarrassed about the unexpected guests, Franklin didn't say. Instead, he noted in a letter to a French friend, he seized the opportunity to satisfy an old curiosity. "Having heard it remarked that drowned flies come back to life in the sun," he wrote, "I proposed making the experiment upon these." He strained out the flies and laid them out on the sieve in the sun. Within three hours, two of the stowaways stirred, stood up, wiped their eyes, and flew off into the London sky. The third fly stayed dead until the sun went down, and Franklin threw it away.

Franklin was encouraged by this result. "I wish it were possible," he wrote, "to invent a method of embalming drowned persons, in such a manner that they might be recalled to life by the solar warmth." The flies had traveled not only across the Atlantic, but through time, and Franklin lamented that he lived "too near the infancy of science to see such an art brought in our time to its perfection." But even before Franklin's parlor experiment, scientists had begun to consider this possibility. In the seventeenth century, Henry Power, a British doctor and an enthusiast of the newly invented microscope, froze some vinegar eels—minute nematodes that live in fermenting vinegar—in salt water. He thawed them, put them on a slide, and determined that they were as alive as before he'd frozen them. Cold, Power concluded, did not have the "killing properties" associated with heat.

Indeed, early thermal explorers such as Robert Boyle (of Boyle's law) wanted to parse heat from cold and concluded that they really couldn't: cold was an absence—of heat, of movement, of energy. In 1848, William Thomson, aka Lord Kelvin, proposed to measure the absence by establishing zero degrees as the point at which no more heat could be lost from one body and transferred to another. Or, to put it in a way that would have made sense to a time-tinkerer like Franklin (who first proposed daylight saving and who wished to come back in a hundred years to see how his country's first century had gone), extreme cold stops the clock. At absolute zero, nothing happens, not even entropy, the loss of heat and order that for physicists marks the direction in which time marches. Which means that bodies frozen and then brought back to life have traveled through time.

But people can't just climb into a very cold freezer and wait for the future. The problem is water. When it freezes, it expands, and inside the body, where the fluid between cells is mostly water, this is a disaster, the same disaster that turns a frozen stalk of celery into a pile of green mush. Individual cells are squashed by ice crystals like middle-seat airline passengers between two sumo wrestlers, until tissues and organs reach what physicists call a "thermodynamically stable configuration"—frozen and unchanging but ruined.

In 1938, Basile Luyet, a Swiss-born, Yale-trained Jesuit priest and biologist who was interested for both theological and scientific reasons in what exactly constituted the difference between life and death, figured out that a liquid like water, if exposed rapidly to very low temperatures, actually gets too cold to crystallize. Instead, it becomes syrupy and inert, its atoms stopped in their tracks, its cellular structure frozen in time like a snapshot—or, as the physicists put it, vitrified—without being destroyed. "One and the same organism," Luyet theorized in his monograph *Life and Death at Low Temperatures,* "may therefore possess a zone of lethal temperatures above zero, a sharp death point slightly below zero, and a zone of nonlethal temperatures some hundred degrees below zero."

The implication was clear: if you could figure out how to cool a living being rapidly enough, you could vault beyond the death zone and into a state of suspended animation. And Luyet did just that with specimens—onion skin, lettuce leaves, chicken hearts—that he dunked into liquid nitrogen. At −170 degrees Celsius, Luyet said, the specimens were "like a watch that has unwound." When he rewarmed the specimens, they picked up their lives where they had left off, and time started ticking again.

This method was limited to minute slices of tissue. The cold of the nitrogen couldn't penetrate more than 1/100 inch quickly enough to leap over the lethal point. While Luyet's solution—dehydrate the tissue as it is cooled, thus further minimizing the crystallization problem—works well for instant coffee and backpacking vittles, he never managed to freeze-dry and rehydrate a dog, in what he called his "pet project." He couldn't even freeze sperm in a way that kept it viable, but in 1949, Christopher Polge, a British scientist, was surprised when semen he'd frozen for an experiment emerged from the freezer with its spermatozoa intact. His experiment, which depended on destroying the sperm, was ruined, and Polge, searching for the culprit, discovered that a bottle of what he thought was glycerin had been mislabeled and was instead glycerol. Further experiments proved that glycerol could stop the formation of ice crystals as temperatures dropped. Within a few years, with the help of glycerol and other cryoprotectants, human sperm and ova were frozen and preserved, and the age of artificial insemination and in vitro fertilization began.

Some scientists remained interested in the uses of low temperature at the other end of life. One of them, Michigan math and physics professor Robert C. W. Ettinger, announced to anyone who would listen that all that stood in the way of our immortality was the perfection of freezing technologies. In *The Prospect of Immortality,* which he published in 1964, Ettinger urged a national mobilization, not unlike the space program, to leap beyond those difficulties and into the limitless future, when medicine will have found a cure for all our ills and the abundance of life will have made for an economy of surfeit. Across the country, small groups formed around Ettinger's manifesto and another tract that appeared at the same time, Evan Cooper's *Immortality: Physically, Scientifically, NOW.* In addition to trying to drum up public support and scientific interest in the field they were now calling *cryonics,* the immortalists began to cast about for a pioneer to head out into the frozen frontier.

They had a few near-hits. In 1965, Wilma Jean McLaughlin's husband agreed to preserve his wife (who was unconscious at the time), but he couldn't find a minister, let alone a doctor or hospital, in his hometown of Springfield, Ohio, to cooperate, and the attempt was aborted. Dandridge Cole, a scientist, said he wanted to be cryopreserved, but he was only forty-four when he had a fatal heart attack in 1965—too young to have made the necessary arrangements. And, as Ettinger later wrote, "In the end the family did what was to be expected—nothing." The following year a woman was frozen, but not before she had been embalmed with the undertaker's usual formalin-based fluid, which, its other attributes aside, makes a lousy cryoprotectant. "There is little or no thought," Evan Cooper wrote in his newsletter, *Freeze-Wait-Reanimate,* "that this first frozen pioneer will rise again."

Finally, in 1966, James Bedford, a retired professor of industrial psychology from Glendale, California, and an active member of the Cryonics Society of California, learned that he had terminal cancer and volunteered to be a cryonaut. Arrangements were made, the family brought on board, a nursing home located that would allow on-site freezing just after death, a cryoprotective cocktail formulated with glycerol and dimethyl sulfoxide (DMSO, a compound that could carry glycerol deep into cells without damaging them), and equipment rounded up, including a Westinghouse iron heart, an early CPR device that would keep Bedford's blood moving after his death and prevent ischemic damage to his brain.

Bedford cooperated fully, except for one thing. He died earlier than anyone expected, in January 1967. But the owners of the nursing home surrounded his head in ice and massaged his heart until the team could assemble at his bedside, hook up the iron heart, perfuse him with the DMSO solution, and whisk him away, wrapped in a quilt and dry ice, in the bed of a pickup truck—a process later documented in *Life* magazine. He was taken to his house, packed into a shipping crate, and then shuttled by pickup truck and station wagon (with soaped windows) across the L.A. Basin for four days, during which his keepers moved him from house to house, eluding coroners and reporters and a hysterical wife who objected to having human remains in her garage; infused him with dry ice in an emergency procedure carried out in a Topanga Canyon park; and finally turned him over to morticians, who shipped him to Cryocare Equipment in Phoenix. There he was placed in a Cryocapsule, a custom-designed vacuum flask filled with liquid nitrogen that sealed in the −196-degree Celsius cold in the same way that a thermos bottle keeps chicken soup hot.

Bedford's strange odyssey didn't stop there. Like a ghost unable to rest, he traveled for another twenty-four years—from Phoenix back to southern California, to the Bay area and then south again to the L.A. Basin, where he spent five years in a self-storage unit, his family topping up his leaky Cryocapsule with nitrogen every month or so, and then a few years each at two separate cryonics facilities before moving in 1991 back to Phoenix, where he remains today. His wife has died, his once-loyal son has abandoned him and called for his cremation, and his trust fund has been exhausted. He's been decanted twice into improved flasks and been peered at and prodded with needles and even a chisel; photos of his transfers circulated from hand to hand and, more recently, on the Web, with cryonicists parsing the meaning of his discolored and distended skin, the puncture marks in his neck, the ice on his pelvis and the blood, still red, around his mouth and nose. They have fought over who should host him and whether he was really preserved properly—it appears that the cryoprotectant didn't have as much glycerol as advertised and, even worse, that it wasn't perfused into his organs but merely injected into his blood vessels. But at least he escaped the fate of six other cryonauts of the late sixties who were abandoned by a swindler and left to rot.

We won't know how grateful Bedford will be for all this attention until he's revived, but in the meantime, a whole cryonics infrastructure has evolved. It includes a biotech startup whose stated business goal is to develop technologies for freezing and storing human organs, but whose top scientists just happen to be committed cryonicists; a company in Florida, Suspended Animations, Inc., that promises to attend to your needs from the time you deanimate until you can be placed in the flask (known in the industry as a *dewar*); insurance policies that will pay your cyronics tab; and two organizations that promise to preserve you, place you in a dewar, care for you in perpetuity, and manage your wealth until you need it again. At last count, there were more than a thousand people signed up for cryopreservation and 157 people—and 40 pets—already in suspension, their bodies vitrified, and their tickets to the future worn on charms that dangle from their wrists and necks.

"I GUESS THIS IS A GOOD TIME to psychoanalyze me, huh?" Jim Gibbons says.

We're smack in the middle of the strait between Harstine and Stretch now, and the storm is upon us. It's a squall of decidedly nonepic proportion—a little pelting rain, a raw, cold breeze, a few whitecaps—but it bears just enough evidence of nature's indifference to make you feel forsaken and frightened. And a little foolish, especially when you notice that a waterway normally choked with pleasure craft is empty, that pleasure boater and commercial sailor alike have evidently had the good sense to stay home today.

"Well, as a matter of fact, I've been wondering something," I answer. "Do you think that knowing that you might not be dead forever makes you less risk-averse than the rest of us?"

"Nah," Gibbons says, without even a pause. "I've been like this all my life. Besides, I think my way is less risky. I mean, when you think about it. Dying, getting buried or cremated or whatever—all the risk is there. Oblivion. That's the risk."

Which is exactly the kind of reasoning you hear again and again at the annual conference of the Alcor Life Extension Foundation in Scottsdale, Arizona, where I initially met Gibbons. The first thing people want to know when they spy a stranger—and there are precious few of us here among the 150 or so attendees—is whether you're a *member* or just thinking about *signing up*. It's only a little creepy; they're not exactly evangelizing. But like many cultists, they are brimming with good news—you have a shot at immortality and not the eternal sojourn in Hamlet's far country or the Judeo-Christian afterlife that Freud called the "nursemaid's lullaby" or the Talking Heads' "place where nothing ever happens," but more, infinitely more, of what you already have: an ever-afterlife worth living.

Like Gnostics everywhere, the Alcor members have a hard time understanding how anyone can resist their secret knowledge, its unassailable logic. It is, after all, a variant of Pascal's Wager, the hard-to-refute postulate that it is always a safer bet to live your life as if God exists and presides over your eternal destiny, even if you think otherwise. Given the stakes, said Pascal, a prudent person always seizes the opportunity to cheat eternal death.

And if logic is not comfort enough, there's always the tour of the Alcor facility, a blocky concrete building shimmering in the Arizona sun amid the malls and the condos and the asphalt. Here, a member can see all the arrangements that have already been made to guarantee his or her trip beyond the death zone, such as the new ambulance sitting in the parking lot, ready to be dispatched as far as a thousand miles away to start the process as soon as possible. (In a few hospitals, Alcor has obtained permission for the readiness team to attend your death, place you in a basket full of ice, insert the cannulas for coolant infusion in your body, and wheel you out on a gurney with the sheet over your head and the Thumper pounding your chest.)

Inside, a picture of a pre-deanimated James Bedford has pride of place, his large-eared, sad-eyed visage presiding over the entrance to the surgical wing. Before you get to the sterile rooms where members get cooled and perfused, you pass by a window that gives you a full view of the eleven eight-foot-tall dewars, their stainless steel shining in the glare of fluorescent lights. Each tank holds four patients or, in the event that the patient went for "cephalic isolation" and "neuropreservation," fifty-five heads. (Alcor's informational package explains, "The spiritual status of cryonics patients is the same as frozen human embryos, or unconscious medical patients.") Because of the specific gravity of liquid nitrogen, the heads don't so much bob like apples as sink like stones. Bedford is here, of course, as are Ted Williams and Ted's son John Henry, although no one will say whether they are suspended together. Walt Disney is not here, nor, according to his biographers, is he frozen under

Disneyland, reports of his cryonic suspension having evidently been greatly exaggerated.

The surgery is as blindingly lit as the sun-bitten desert outside. Sound clangs off the hard white walls, turning our tour guide's narration into an echoing litany of gruesome details. We've seen the table where the perfusion takes place, the medical devices—essentially, a heart-lung machine—that flow the coolant through the body, and the jig that holds the head of the neuropreserved while the Alcor staff ministers to it.

"Nature created a very good encasement for your brain," the guide says, explaining why Alcor abandoned its attempts to remove brains for storage. Next to the jig, which looks like a complicated globe stand, is the surgeon who, the guide tells us, performs the cephalic isolations.

I ask him which blood vessels are used for the perfusion and how they are tied to the equipment. He says, "I don't know. I just do the decapitation and leave."

THE TOUR CAPS A THREE-DAY CONFERENCE, held at a nearby hotel, that promised an "Inside Look at the Science and Medicine of Tomorrow." Which will, apparently, be a time when tiny nano-robots will voyage like molecular versions of Raquel Welch into your diseased organs and repair them from the molecules up, when healthy lives will last so long, maybe indefinitely long, that people will have many careers and families and century-long vacations, and when coolant technology will allow for perfect preservation. We've heard from the scientists about the triumphs of cryopreservation, such as the new high-tech coolant called M-22 (named for the temperature [-22 degrees Celsius] at which it vitrifies organs), seen photos of brain slices pre- and postvitrification—just a little bit of damage, the scientist says—and the rabbit kidney that was removed, vitrified, and then placed back into the rabbit, who looks alive, if not quite ready to hop away for a little tinkle. We've learned how to talk to our legislators about making the world a safer place for cyronics—some states are evidently squeamish about the idea, others merely befuddled about whether it falls under medical or mortuary regulations—and how to preserve our wealth so that we can take it with us. We've heard the origin tales of cryonics, including a photo-illustrated version of Bedford's postmortem odyssey, from the founders themselves; have lunched with the luminaries; and have meeted and greeted under the desert night sky.

Alcor is part of a thriving life extension subculture. "I'm taking a hundred fifty supplements a day," one woman told me by way of small talk, and I also heard about the latest in vitamin preparations, high colonics, caloric restriction (scientifically speaking, reducing your food intake is the single most effective way to extend your life, at least if you're a lab mouse), and other ways of living well, dying extremely old if at all, and leaving behind a suspended body if you do. "Cryonics is plan B," said the supplement woman.

The hotel swimming pool stays empty and the presentation rooms full, people riveted to the brain slices and the nanobots, even though these scientists are clearly giving their usual stump speeches and nearly everyone has heard all of this before. They're nodding to the PowerPoint, all but mouthing the words of the lectures like fans singing along with their favorite band, and they know those founding myths and scientific verities as well as any born-again Christian knows his Genesis and his Beatitudes. They've been paying very close attention to their prospective futures.

Which is exactly what Aubrey de Grey thinks we all should be doing. Rail-thin (although not a caloric restrictor) with sharp features and a beard that looks like a hedgehog clamped onto his chin, de Grey, forty-five, is a rock star in the life extension world. Over a crack-of-noon beer in the hotel bar, he's lamenting the fact that so few people are compelled by the logic of the immortalists. It's basic psychology that stops us, he thinks.

"We need to put aging and death out of our minds so that we can get on with our miserably short lives and not be preoccupied by it all the time." This sort of "psychological self-management," he says, made sense when there really was nothing to be done, but with advances in science like the ones we've been hearing about, "We're in a completely different situation, one where the learned helplessness is a big part of the problem." We've fallen into a "pro-aging trance," he says, and that's why so few people sign up for cryonics and, even worse, why we treat aging and death as existential inevitabilities when they are, in reality, an urgent public health crisis.

De Grey talks extremely fast, and in the din of the bar, his British-accented, beard-muffled rat-a-tat turns into a slurry from which words about "mitochondria" and "extracellular protein links" occasionally emerge. He's trying to explain his program—Strategies of Selectively Engineered Negligible Senescence (SENS), an approach that reverse-engineers "age-related dysfunction," boils it down to its seven constituent parts, and suggests programs for addressing each one—but most of what I'm getting is his urgency, as prophetic as his beard and his rush to speak, to "get people to snap out of the trance." De Grey recently left his day job as a computer scientist at Cambridge University to become the chairman and the chief scientific officer of the Methuselah Foundation, where he coordinates the attempt to put the principles of SENS into action. He's raised more than $8 million, half of it for the Methuselah Mouse Prizes, which are offered to scientists who can increase the life span and decrease the signs of aging of the common lab mouse. De Grey thinks that the MPrize will lead to dramatic results that will in turn "make people more inclined to agitate for a solution to the technical problems." Because once you get past this irrational resistance, the immortalists think, mortality is nothing *but* technical problems.

Take the Hayflick limit, for instance. In 1961, Leonard Hayflick discovered that cells can only divide about fifty times before they stop replicating. Hayflick, himself apparently stuck in the pro-aging trance, took this to be the sign that mortality is inborn, that we are programmed at the cellular level to decay and die. But it turns out that the reason for the Hayflick limit is not so much teleological as physiological: at the end of our chromosomes is a stretch of DNA called a telomere. With each replication, the telomere grows shorter until it is gone, at which point your chromosome, like a shoelace after the plastic doohickey at the end falls off, starts to unravel, with disastrous results for the cell replication that keeps us healthy.

It turns out, however, that certain cells are not subject to the Hayflick limit—notably, stem cells (which is why they are so promising for medicine) or cancer cells (which is why they kill us). What those cells have that normal cells don't is telomerase, an enzyme that regenerates the telomere after it divides. In 1998, two scientists introduced telomerase, via a virus, into normal human cells taken from retinal and foreskin tissue. The experiment succeeded in immortalizing those cells, and whatever the ultimate value of an immortal foreskin, it proved that cellular decay is no more natural or inevitable than tooth decay. And although prevention is more complex than brushing and flossing, there is no reason why normal cells (and the organisms they constitute) can't be as immortal as cancer cells—still susceptible to poisoning and burning and other destructive insults, but otherwise capable of living forever. Death, in other words, is not built into life.

Indeed, says de Grey, to believe that there is such a thing as a normal human life span is simply ignorant, based on a mistaken idea of what is "natural." "That idea does help make it look like mortality is out of our control and that aging is the way things ought to be," he told me. But what would the state of scientific medicine be if we simply accepted life as it presents itself on the grounds that this is what nature intended? In this respect, all attempts to cure disease are "unnatural"; killing bacteria with antibiotics is just as much an affront as using telomerase to overcome the Hayflick limit. And where ignorance once led us to think of infection as the will of God, so, too, it now makes us think that mortality is an immutable condition rather than just another disease to be cured. There is no reason to continue to accept dying as our lot, according to de Grey—not even the fact that an indefinitely long life, even one free of the ravages of aging, is bound to be full of other kinds of suffering.

"I've had heartbreak and setbacks, and life is still pretty damn good fun," he says, wiping beer froth from his moustache. "Now maybe I'm not typical, but if heartbreak is so bad, then everybody ought to kill themselves before they are heartbroken. While we're at it, I have an idea. Let's blow up the whole world and save everyone right now."

IMMORTALISTS ARE LIKE ESPERANTISTS or devotees of Ayn Rand, shaking their heads at the persistence of irrationality in the face of reason, at the ignorant who cling to the mess and suffering of life-as-it-is when an alternative is available. They shake their heads at the pigheadedness of a legal system that forbids them to get preserved before the actual moment of death; the 1992 California decision prohibiting the premortem freezing of Thomas Donaldson, who wanted to go into suspension before his cancer destroyed his brain, is the cryonicists' Dred Scott decision, the prevailing of prejudice against fairness and common sense. And when the veil of ignorance seems to be lifting, they are cheered. As they are when Aubrey de Grey introduces a ten-year-old girl, Avianna Vyff, and presents her with a special award—not only because Avianna, inspired by a diabetes fund-raiser, has raised $3,000 for the MPrize by going door-to-door and explaining to her Austin neighbors that getting old and dying is a bad thing and efforts to prevent it a worthy cause, but also because this child, who hasn't yet fallen into the pro-aging trance, represents the best hope for the future of immortality. Of course, Avianna may be in a different kind of trance: her mother, Shannon, a leggy blonde whose entry into the room causes a stir, perhaps because of her very, very short dress, wrote *21st Century Kids,* a children's novel narrated by a girl named Avianna, who, along with her brother, is killed in a car crash, cryonically suspended, and then revived in 2189.

In Vyff's version, the future is not without its difficulties: an
irreparably damaged environment, as well as some Mad Max–like
political chaos. But there are nanoshields against the polluted air
and nanobots to do people's bidding, and citizenship has been
granted to dolphins. This optimism is an integral part of the cryonics dream because, as you hear repeatedly, the society that is
capable of thawing the suspended will necessarily be advanced
enough to heal them, which is to say that you will wake up in a
medicotechnological paradise. It's a little infectious, this goofy
optimism, this unbounded faith in the trajectory of progress,
and it makes you wonder whether maybe your skepticism is just
reflexive pessimism or, worse, envy of these people who don't seem
to think that technology will always have a dark side or that life,
especially a life rid of mortality, has to be a raw deal. And as Rudi
Hoffman—the guy who shows up at the conference every day
in a different Western-style shirt, each one with "May I help you
with your cryonics insurance needs?" embroidered on the back—
explains how easy it is to afford one of his policies, you have to
grant that once you're dead, it really doesn't matter whether
you're moldering in the ground, scattered on the wind, or bathing in liquid nitrogen on the remote chance that you will wake
up like Franklin's flies. What holds you back besides your disgust,
the least reliable of our moral sentiments and the one most likely
to turn out to be fear of novelty masquerading as righteous belief
in the old ways?

So you don't speak your most churlish thoughts. You don't
ask them how they know that the society of the future won't be
as morally backward as our own or even worse, that people won't
decant the frozen for use as slaves or for target practice. You don't
point out that more than anything else, the immortalists seem nos-
talgic for the future, pining for an epoch from which time bars
them, resentful of their unborn descendants who might not die
from what will surely kill those of us unlucky enough to live now.
And when the woman from Tucson, the one who, within five
minutes of meeting you, tells you cheerfully about her husband

who is "suspended," about the movie she made of his cephalic isolation, and of her visits to his tank, waves her hand across the crowd gathered on the Marriott patio under the huge desert moon, taking in all of these affluent, white-wine-and-cheese-grazing navel gazers who count their days in supplement spoons, and says, "You know the best part? The best part is that when we wake up, we'll all wake up together," you don't quote Wallace Stevens on the complacencies of the peignoir or object that the future is not a gated community. Because you have to admit that you just don't know whether that's true.

JIM GIBBONS AND I have run out of things to say to each other. We're alone with our thoughts and the sound of the rain. I'm too cold and sick at heart to ask any more questions. Not to mention that my notepad is soaked, my anxiety has rendered my memory unreliable, and I left my recorder back at the dock. I'm hoping that when they find it and play back my last interview and hear my seduce-and-betray tactics, they won't think I'm too much of a jerk.

My cell phone rings. I dig it out from under all my layers. It's my wife. I click off the phone so that I won't have to lie to her. If it gets really dire, I'll call her back. I try to imagine that conversation and the one with my son, but guilt stops me. All I can think of is how strange it will be, immediacy and infinite distance at the same moment. I think of climbers dying on Mt. Everest, saying good-bye from their satellite phones, how they must have wished they could phone themselves home, ride the stream of electrons carrying their voices up into space and down into their own warm kitchens.

The immortalists have thought about this, too, about making your consciousness portable. It's an important subject because even the best cryopreservation involves extensive brain damage—ischemia inevitably sets in between deanimation and immersion, and M-22, its antifreeze properties aside, is pretty toxic. Alcor acknowledges publicly that the chances of actually preserving your brain without damaging it, given current technology, are very slim. More likely, you'll have to stay in suspension until future scientists figure out how to restore a devastated brain—with nanosurgery, perhaps, or maybe with brain implants, which will replace neurons and synapses with silicon circuits, or maybe even by digitizing everything stored in your neural network and uploading it to the Internet. Your brain might be ruined, your heart stilled, your body putrefying, but if the information encoded in your brain can be preserved, you are still not dead. You might wake up as a routine in a computer, as a digitized self, Descartes' dualism brought forward into the twenty-first century—not as the soul extracted from the body, but as the information distilled from the meat.

If your memory is spotty, your personality disrupted, and all the familiar pathways wiped out, then it may well be that nothing connects you to the you that you were before you got suspended, that you will wake up as someone else entirely. You will have suffered what cryonicists call "information-theoretic death," the one kind of death that is, in their view anyway, absolute. But on the other hand, according to bioethicist Jay Hughes, maybe you won't be dead at all. Or, to put it better, maybe death itself will be made irrelevant by technologies like cryonics and whatever advances medicine makes that allow doctors to revive the frozen. Hughes is a leading light among transhumanists—people who embrace the idea that our destiny is to reengineer ourselves as a new species—and thinks that these same developments will render cryonics unnecessary at exactly the same moment that it becomes plausible; you won't need to get frozen because you'll be able to be cured in the present. While he thinks that information-theoretic death is some kind of absolute horizon, Hughes believes that this is important only to the extent that we live as single identities in the first place.

"The continuity of the self is an illusion," Hughes said, "a probabilistic continuity between one state of mind and the next, in which the changes are minor enough that you still consider yourself the same person." Technologies such as cryonics are bound to weaken the grip this illusion has on us.

We are already grappling with the possibility that a person can have two distinct identities, he pointed out: an Alzheimer's patient, for instance, in whom all traces of identity have disappeared, is a chimera when it comes to law and ethics. For instance, Hughes wondered, should we honor a do-not-resuscitate order in the advance directive of an Alzheimer's patient? Is the person who made that order the person on whom it will be carried out? Perhaps the self emergent in the new neurochemistry of that person would be happy to live forever. On the other hand, perhaps the pre-Alzheimer's self must think about stewarding the life of the self that will emerge as the neural fibers get tangled. Bioethicists call these conundrums "Odysseus questions," referring to the way Odysseus lashed himself to the mast and ordered his men to ignore any commands he issued under the spell of the sirens. But what if Odysseus didn't have to remain the man he was? What if he was free to live many identities? What if the thing that held him together was not some biological necessity but his immersion in a story about being an individual, and what if his own private fiction changed?

Hughes added that there are other prominent examples of people living discontinuous lives: born-again Christians, for instance, who die to be reborn. But most resurrections require sacrifice, and if individual identity is the cost of immortality, if the "death of death," as Hughes calls it, hinges on this blasting away of the self, then a technology like cryonics, which challenges the fiction—powerful, useful, sometimes even noble—that we are single beings, one to a body, may succeed in a way that no one expects.

"Ultimately, I think people have to get over their hang-up about identity," he told me, adding that the whole question of whether we are one or many will eventually become incoherent. "As we live longer and longer and have more and more opportunities to modify our brains, we're going to see that as a pointless philosophical debate." Some cryonicists believe that this is the development they're going into cold storage to await, that they won't wake up as thawed-out-and-healed versions of the bodies they used to have but as data awaiting implantation into new bodies, each one with a new cerebral architecture, its characteristics chosen like automotive options—a different sex perhaps or maybe even the ability to fly.

OR BREATHE WATER. That's what I'm wishing for at this moment—gills. That, or a thick blubbery coat like the one on the harbor seal that's been circling the boat for a while, as if sticking around to see how it all turns out. In fact, right now I can think of a million improvements on this big-brained, thin-skinned, land-bound body of mine, all of which will no doubt be available in 2189. I'm festering with resentment of my descendants, drowning in nostalgia for the future.

Gibbons is wondering aloud about the boat that is coming to fetch us—a seventeen-foot skiff, he says, that may have a smaller motor than ours. He's not sure that it will be able to tow us all the way back to the marina. I picture the two stranded boats tied together, one pulling the other under, all hands lost, information-theoretically and otherwise.

"I wonder if I should call Tobin," he says.

"Tobin?"

"One of my foremen. He's harvesting off Harstine, in the big boat."

"Big boat?"

"The *Soha*. Thirty-four footer."

"Where did you say Tobin was?"

Gibbons points to the orange windsock, now a distant bright spot against the steely gray sky. "If we'd made it around that point, we'd have seen him."

"You mean when we broke down, he was, like, five minutes away?"

"More or less."

"Yeah, Jim, I think this would be a good time to call Tobin."

It's a little complicated—calling the office, quarreling with his manager about how urgently the boat is needed, whether it's worth losing the last hour or so of harvest, getting Tobin's number, ringing his phone until he hears it over the *Soha*'s machinery—but he eventually persuades Tobin to come and fetch us. "Sooner than better" he says, and for a moment it seems that Gibbons might actually be anxious. ("It's getting pretty bad out here," he says, the first indication that he's actually noticed.)

But soon the rain stops, the whitecaps subside, and we seem to be drifting straight west to a closer shore than Stretch. Tobin is on his way and relief washes crisis out of the boat. We take down our hoods and I remove my Mae West. Fifteen minutes later, we've drifted so close to shore that we have to drop anchor so we won't go aground. The *Soha* and the skiff arrive at the same time. Gibbons sends the skiff home, while Tobin deftly guides the *Soha* abeam, fastens the tow rope, gives his boss the requisite razzing, and welcomes us aboard.

No way that would have towed you back," Tobin says later. "The aluminum on that thing is only this thick, probably would have split in two."

On the run back to the marina, I call my wife, and we trade disaster stories: while I was contemplating mortality on the Puget, a vicious Nor'easter was blowing two trees across our driveway, knocking out the power, and flooding our basement, and she was calling to say that she was headed for some friend's house to stay the night. I hang up and as the *Soha* works its way back down the sound we talk about Tobin's girlfriends, about growing up on Puget Sound, about clams and our kids and the history of the boat, about everything really except peril and mortality, the poor prospects of the human flesh amid all that is arrayed against it.

Not that Gibbons doesn't try to talk about that, or at least about the way he's trying to increase his odds. He tells Tobin why I've come to visit and stops just short of asking Tobin for his opinion of cryonics. But if Tobin has any thoughts about his boss's post-deanimation plans, he isn't saying. Neither is Gibbons's family, at least not to me. After we fish the boat out of the water and haul ourselves back to Jim's house; after a hot shower and a hotter burrito supper and the raspberry pie Mary Gibbons and their seven-year-old son, Ian, whipped up; after soaking in the domesticity long enough for the afternoon's dread to wash off, to begin its transformation from near-catastrophe to mere story, Jim tries to get them to talk to me about cryonics. He hints around, he brings it up directly, he appeals on my behalf, he leaves the room, he hovers, but Lucas, the eleven-year-old, wants to go down the street and visit the neighbor's animals, and Ian has a game he wants to play that involves hurling himself headlong into the couch, and although Mary keeps promising, she keeps finding something else to do. Jim never gets to hear from his family what they think of his plan—something he seems at least as curious about as I am.

Finally, when Jim asks the boys whether they know what will happen if he dies, Ian bites. "I know you're going to get frozen so you can come back to us," he says. I glance over at Mary in the kitchen. She looks as if she wants to say something but turns back to her burritos instead.

Later, while Jim keeps Ian distracted, I settle into the living room with Mary Gibbons, who has an unlined face and a long, thick reddish braid that makes her look younger than her fifty years. "Jim has lots of ideas," she tells me. She says it with only a touch of resignation and with much more affection than rancor. Some of his ideas have turned out pretty well, after all. This is all she says, and what we end up talking about is what she's seen at the bedsides of the grievously ill or injured, about all the uncertainty that haunts those last momentous decisions and how no one in attendance—doctor, patient, family, clergy—can ever really know when to say, "Enough."

Because there can never be enough life. That's why we have medicine in the first place: to stave off suffering and ultimately to forestall death, to provide endless hope, to do, that is, what religion did before science kicked it out of its place as the source of our knowledge about ourselves. And if scientific medicine has replaced the nursemaid's lullaby with the little boy's vision of his once-dead dad striding into his living room, it has also left us with an inexorable logic, its own version of Pascal's Wager: that, given the stakes—which, without the nursemaid, are surely oblivion—if there is something else to try, some other procedure that might work, some faint glimmer of the possibility of more life, it should be done, even if it destroys us. For what is the basis of refusal? Why would anyone say no to the immortalists? Surely not horror at the ruptured boundary between life and death or at the vision of an eternity spent in a gleaming steel bottle of liquid nitrogen or at the hubris and greed of gobbling up tomorrow's wealth as ravenously as yesterday's. Horror is not up to the task of telling us what to do with the knowledge it never would have let us pursue in the first place. Its certainties are no more honestly earned than science's. The truth is harder than that.

AFTERWORD

These stories were collected over the last eight years. During that period, the webs of misunderstanding (and sometimes deceit) spun out of our noble lies have only gotten more tangled. As I finished writing this book at the beginning of 2008, front-page stories appeared in newspapers and magazines about antidepressants and addiction, about brain death and organ transplant, about the boundaries between health and illness and between enhancement and treatment. Here's a quick tour of the current landscape of our confusion.

In January, the *New England Journal of Medicine* reported on the seventy-four clinical trials that have been submitted to the Food and Drug Administration (FDA) for the twelve leading antidepressants. Only thirty-seven were viewed by the FDA as positive. Of the remaining thirty-seven studies, twenty-two were never published and eleven were written up in a way that, as the authors of the study put it, "conveyed a positive outcome," even though the FDA viewed the results as negative or questionable. This means that a person reading *every* published study about the twelve leading antidepressants would come away with the impression that 94 percent of the trials yielded positive results, when in fact only 51 percent did. A Canadian study published at the same time reviewed the effectiveness of Paxil and concluded that it was no better than placebo and had far more side effects.

The depression industry chugged along unimpeded, however, fueled by its brilliant version of the noble lie.

Also in January, the *New York Times* reported on its front page that sales of Pfizer's drug Lyrica, which had been a disappointing performer as a treatment for pain related to diabetes, had increased 50 percent in 2007, to $1.8 billion. The company was expecting an additional 30 percent increase in 2008, in response to a blitz of television ads trumpeting the drug's effectiveness, newly endorsed by the FDA, for fibromyalgia, a notoriously hard-to-treat condition featuring an assortment of vague symptoms such as chronic pain and fatigue. "Today I struggled with my fibromyalgia," says the woman in the commercial (according to the *Times*). She turns to the camera and adds, "Fibromyalgia is a real, widespread pain condition." The evidence for this? That Lyrica makes fibromyalgia better, which means that it must be biochemical in origin—and therefore a real disease. Evidently, the drug companies are still following the advice Dwight Anderson gave to the Research Council on Problems of Alcohol: create the disease and the people will follow.

There is a gopher in Pfizer's field of dreams, however: Dr. Frederick Wolfe, the director of the National Databank for Rheumatic Diseases. In 1990, Wolfe did the original work that established the diagnostic criteria for fibromyalgia, but now he's having second thoughts. "Some of us in those days thought that we had actually identified a disease," he told the *Times*. "Which this clearly is not." Wolfe regretted his error. "To make people ill, to give them an illness," he said, "was the wrong thing." As another rheumatologist put it, fibromyalgia was an iatrogenic illness, something people catch from going to the doctor. "The more [these patients] seem to be around the medical establishment, the sicker they get."

That's pretty much how one of Pfizer's consulting doctors, Dan Clauw, saw it. "What's going to happen with fibromyalgia is going to be the exact thing that happened to depression with Prozac," he said—meaning, of course, that it would come to be seen as a "legitimate problem," not that it would become another fictional disease that owed more to commerce than to medicine. But neither Pfizer nor Dr. Clauw needs to worry about Dr. Wolfe or anyone else revoking fibromyalgia's disease license. Although we have come a long way since Anderson first came up with the idea of inventing diseases to accomplish extra-medical purposes, no one has yet figured out how to uninvent them. (Homosexuality is the happy exception to this rule, although it is always possible that further research into the origins of sexual orientation, coupled with a shift in the political climate, could lead to re-diseasing it.) With fibromyalgia, no less than with depression, the disease idea has too much going for it—profit-seeking drug companies, people who want relief from suffering and doctors who want to provide it, a cultural climate in which any idea bearing the imprimatur of science has credence—to be undone by mere logic and reason.

That's a point that couldn't be lost on Irving Kirsch. He's the psychologist who did the original analysis of the FDA studies and showed that the placebo effect accounts for most of the small improvement attributed to antidepressants in clinical trials. In February, he and his colleagues published an article further crunching the FDA's numbers. The analysis showed that antidepressants' advantage is a function of severity: the worse your depression, the more likely it is to respond to drugs. Even this effect, Kirsch wrote, is a result of the placebo effect: severely depressed patients have a lower placebo response, which makes any medication response show up more strongly in the statistics.

This doesn't necessarily mean that the less severely depressed people taking the drugs are lying or have simply been duped when they say they feel better. It's much more likely that the kind of feeling better they experience doesn't show up robustly in the measurements. It is very difficult, after all, to construct a test for something as idiosyncratic and subjective as a feeling state. As a result, there may well be two kinds of depression out there: the severe and relatively rare kind whose presence and remission register nicely on psychological tests and the much more common malaise—call it pessimism or disappointment—that the existing instruments are no more sensitive to than magnets would be to feathers. It's important to keep in mind that doctors, especially family doctors, who provide most of the antidepressant prescriptions, don't routinely give the diagnostic tests. They ask a few questions and prescribe the drugs, telling the patient that he or she is depressed. So while depression may officially be a condition in which you have five of the nine *DSM-IV* symptoms, unofficially it is something else: as Bunky Jellinek might put it, depression is what your doctor says it is. With so many people walking around with that diagnosis (and feeling better, in whatever ineffable way, on the drugs), it soon seems plausible that there's an epidemic out there, and increasing numbers of people are bound to think that their dejection and worries mean that they've contracted the disease.

A simple reform is possible: allow only severely depressed people into clinical trials and require that patients score high on the SCID (the diagnostic test tied to the *DSM-IV* symptoms, a test that, by the way, takes nearly forty-five minutes to administer) before granting them a prescription. The problem with this—aside from the obvious dent it would put in Big Pharma's profits—is that clinical trials could never be filled with the severely depressed, as they are relatively few in number and unlikely to present themselves at test centers. Even more important, however, there is no good reason to deny people drugs that make them feel better even if they don't have a disease in the first place. It may be worth debating whether it's a good idea to make a casualty of the truth in order to get drugs into the mouths of unhappy people, but it is certain that for now the noble lie about depression is the best way to give people a break. Why so many people seem to need a Prozac break . . . well, that's a question that future historians are bound to puzzle out when this noble lie about our unhappiness gives way to the next.

Another front-page story that appeared in February made it clear that sometimes the stakes of our noble lies are life and death.

The *Times* reported that a doctor in California, Hootan C. Roozrokh, had been indicted for three felonies after an organ harvest went bad. Roozrokh was in charge of a non-heart-beating-cadaver donation, the kind that Nicholas Breach wanted to make, in which the donor is removed from life support in the operating room, is allowed to die, and, after a decent interval, is opened up for organ removal. Ruben Navarro, the patient in Roozrokh's case, didn't die right away, however, and the doctor administered larger doses of morphine and Ativan (both drugs that can hasten death and are of questionable value in the case of a patient who, like Navarro, is severely brain damaged and already unconscious). The doctor also infused Navarro with Betadine, an antiseptic that would benefit only his organs and not the patient himself. As it turned out, none of this was to any avail: in the eight hours it took Navarro to die, his breathing was so compromised that his organs were destroyed by lack of oxygen.

Non-heart-beating cadaver donations "can make some doctors and nurses skittish if they have not previously witnessed one," an expert told the *Times,* and apparently Jennifer Endsley, the nurse who blew the whistle on Roozrokh, was one of the sensitive neophytes. One donation advocate worried that her squeamishness would spread, that as the case hit the news it would "give some support to the myths and misperceptions we spend an inordinate amount of time telling people won't happen." Airing this dispute, in other words, may alert people to the fictions that underlie the transplant industry: that we know exactly what constitutes the moment of death, and that science has established this line sufficiently to assure us that the donor is really dead. Indeed, it's hard to imagine this trial not, at some point, touching on the fact that doctors have invented these boundaries in order to facilitate what would otherwise not be permissible.

Anderson's and Jellinek's noble lie, on the other hand, doesn't seem to be in any jeopardy. In late February, *Newsweek* ran a cover story on the "hunt for an addiction vaccine," an article that was accompanied by the requisite photos of your brain on drugs, as if showing that blood flowing differently in a person who gets high three or four times a day than it does in someone who doesn't somehow proves that addiction is a disease. And, of course, the idea that you could have a vaccine that prevents addiction (as opposed to the much more ominous-sounding *drug that prevents people from behaving in a certain way*) is only further "evidence" for the disease model. Lest there be any doubt about the direction of addiction policy or what it means to call anything a disease in the age of Big Pharma, Nora Volkow, the director of the National Institute on Drug Abuse, told *Newsweek,* "In 10 years we will be treating addiction as a disease, and that means with medicine."

Also in early 2008, college shootings raised the question of whether mental illness is the cause of immoral behavior (and, in one case, of whether antidepressants can lead to violence), and disputes about gay marriage and abortion lurked just under the surface of the presidential campaign. But by far the most spectacular—and bizarre—manifestation of the noble lie was to be found on C-SPAN, which in February broadcast the full day of sworn and contradictory testimony from Roger "the Rocket" Clemens, baseball's greatest active pitcher, and Brian McNamee, ex-cop, personal trainer, and steroid dealer to the stars. Thin and rangy with beady eyes, McNamee played the weasel to Clemens's rat, backing the corn-fed Texan into a corner with his account of providing him with steroids, injecting him (in his Cy Young–awarded buttocks, as we heard over and over) with human growth hormone, and in a move that couldn't help but recall another powerful man undone by the underling who serviced him, keeping the physical evidence—not a blue dress but some bloodied bandages and possibly DNA-containing syringes. McNamee had his scoundrel mojo working: the fact that tattling on Clemens was tattling on himself only increased his credibility. The sole defense that the lip-licking Clemens could muster was the fact that he was Roger Clemens, and so he couldn't have done anything bad—an argument that recalled George W. Bush's insistence that because the United States doesn't commit torture, waterboarding can't be torture.

But the real spectacle wasn't to be found in Clemens's backing himself into the perjury corner—a common-enough outcome, after all, when a public figure finds himself on the verge of disgrace. It was in the fact that this tawdry he said/he said took place in the U.S. Congress, which presumably has a country to run, a war to worry about, and an economy to salvage from an *Exxon Valdez*–size shipwreck. And not only that, but the debate quickly broke down along party lines. The Republicans came out for Clemens and the Democrats for McNamee, the former pointing out that the weasel was, well, a weasel, and the latter that the Rocket had crashed back to earth, disappointing the millions of kids who looked up to him as a role model.

This partisanship was a little hard to understand, at least at first (although after nearly eight years, it has probably become a reflex for Republicans to rally behind a Texan who has a sense of entitlement the size of his home state, who leaps from cliché to cliché as if they were so many islands in a vast sea of inarticulacy, and who evidently thinks that repeating the same not-so-noble lies over and over will transmute them into the truth). But then you had to notice that the Republicans, at least implicitly, were standing up for liberty—the freedom to do whatever it takes to gain advantage—and the Democrats for a level playing field, which is, of course, the same dream that has fed affirmative action, busing for integrated schools, and the other social justice initiatives associated with that party. Clemens's case, in other words, fell into one of the deep fissures in American public life: the tension between freedom and self-restraint, between the idea that our lives ought to be dedicated to the relentless and unbounded pursuit of success and the notion that the invisible hand is much better at urging us on than at showing us where to stop.

You wouldn't want to make too much of the way the Clemens hearing fractured along party lines. That could have as much to do with campaign contributions as anything else, and it could as easily have broken the other way. It was, remember, a Republican president's bioethics commission that came out against steroids and other enhancement technologies. And it's not as if the Republicans mounted a spirited defense of Clemens's drug use; indeed, that was the last thing they wanted to talk about. Turning great and perplexing problems into petty partisan squabbling is what our politicians do best—often, it seems, in hopes of rallying their base. There's nothing like an insoluble problem to generate empty rhetoric, while simultaneously keeping people's minds off problems that can be solved.

In this case, the bad faith that underlies the politicians' posturing is the same bad faith that gives rise to the noble lies collected here: the claim to know exactly what nature is and what kind of people it intends us to be. The Democrats (in this case) were making hay of the contention that Clemens had not simply *cheated* by breaking Major League Baseball's rules but that he had done something Mother Nature had not meant him to do. (It's too bad Clemens didn't just fess up and then turn the tables, asking his inquisitors which of them had used Viagra or alcohol the night before or Prozac or caffeine that morning.) The Republicans, on the other hand, held forth as if taking steroids suddenly were of no more importance than corking a bat. Representatives on both sides declaimed as if they knew where the bright line between cheating and blasphemy lay, and here they were doing precisely what any self-promoter must do: provoking the anxieties of an audience and then offering to settle them.

Because we are all worried about what will become of us now that we've begun to crack nature's code. Leon Kass has written that "human nature is on the operating table," and science, which has done so much to put us there, simply can't tell us what to do next, no matter how much we might like it to. We can't diagnose away our problems. Even if they are cleverly fashioned, our noble lies can't last. As Ahab reminded Starbuck when the crewman told the captain that his relentless pursuit of the whale was blasphemous, "Truth hath no confines." It will always overflow whatever puny categories we manufacture for it. No matter how noble our lies are, the truth will always reemerge, and then we will have to fashion new ones.

NOTES

INTRODUCTION

1 *In the winter of 1816*   Laennec tells this story in *De l'auscultation Médiate.* An 1821 translation—J. Forbes, trans., *A Treatise on the Diseases of the Chest*—was reissued by the Classics of Medicine Library. Laennec is also the subject of a chapter in S. Nuland, *Doctors: The Biography of Medicine* (New York: Alfred A. Knopf, 1988), pp. 200–237. See also A. Roguin, René-Théophile Laennec: The man behind the stethoscope, *Clinical Medicine & Research* 4(3) (2006): 230–235.

2 *"to his habits"*   Hippocrates, "Of the Epidemics," book 3, section III, from F. Adams, trans., *Hippocratic Writings* (Chicago: Encyclopedia Britannica, 1952), p. 49.

3 *But even this low-tech approach*   See Nuland, pp. 3–30, for an account of the ascent of the Hippocratics over the Aesculapians.

3 *There is a painting*   *Laennec and the Stethoscope* was painted by Robert Thom as part of the Great Moments in Medicine series sponsored by Parke, Davis, and Company. These paintings originally appeared as inserts in *Modern Pharmacy,* Parke's house organ, but were so popular that they were reprinted to hang in doctors' offices and drugstore windows. A copy of Thom's *Laennec* hangs in the Laennec Museum in Nantes, Laennec's hometown. See J. Duffin and A. Li, Great Moments: Parke, Davis and Company and the creation of medical art, *Isis* 86(1) (1995): 1–29.

7   *"found in a state of"*   Egerton Y. Davis, letter, *Philadelphia Medical News* 45 (1884): 673. The letter is reprinted in S. W. Bondurant and S. C. Cappannari, Penis captivus: fact or fancy? *Medical Aspects of Human Sexuality* 5(5) (1971): 224–233.

7   *"When I arrived"*   Davis, p. 673.

7   *"As an instance of Iago's"*   Ibid., p. 673.

8   *"the insertion of a"*   J. S. Oliven, *Sexual Hygiene and Pathology* (Philadelphia: J. B. Lippincott, 1955). Cited in E. F. Nation, William Osler on penis captivus and other urologic topics, *Urology* 2(4) (1973): 468–470.

8   *"Hays, for Heaven's sake"*   H. Cushing, *The Life of Sir William Osler,* vol. 1 (London: Oxford University, 1925), pp. 240–241. See also Bondurant and Cappannari, p. 232.

8   *Egerton Y. Davis continued*   A list of Davis's entire corpus can be found in the *Osler Library Newsletter* 38(1) (October 1981): 1–6, www.mcgill.ca/files/osler-library/No38October1981.pdf, retrieved July 19, 2007.

8   *But he shouldn't have been surprised*   Nation, p. 469; Bondurant and Cappannari; and B. Musgrave, Penis captivus has occurred, *British Medical Journal* 280(6206) (1980): 51. Notably, Musgrave does not cite the Davis case, even as he insists that Osler's invention really existed.

9   *"common yet under recognized disorder"*   GlaxoSmithKline press release, June 10, 2003, p. 1, www.gsk.com/press_archive/press2003/press_06102003.htm, retrieved July 21, 2007.

9   *"can produce severe insomnia"*   Ibid., p. 1.

9   *"17 percent of adults"*   Ibid., p. 2. According to a subsequent study, these reports were greatly exaggerated, mostly because anyone who complained of *any* of the symptoms was counted as an RLS patient. The true prevalence was 2.7 percent. See R. P. Allen et al., Restless legs syndrome prevalence and impact: REST general population study, *Archives of Internal Medicine* 165(11) (2005): 1286–1292. Even those numbers are skewed by a method that oversampled people with RLS. See S. Woloshin and L. M. Schwartz, Giving legs to restless legs: a case study of how the media helps make people sick, *PLoS Medicine* 3(4) (2006): 452–455.

9   *"Individuals with RLS"* GlaxoSmithKline press release, p. 1.

10   *It was as if the fact*   N. Wade, Scientists find genetic link for a disorder (next, respect?), *New York Times,* July 19, 2007, p. A20.

11   *With $2 trillion at stake*   Government Accounting Office, *Health Care Spending: Public Payers Face Burden of Entitlement Program Growth While All Payers Face Rising Prices and Increasing Use of Services* (2007), p. 7, www.gao.gov/new.items/d07497t.pdf, retrieved December 5, 2007.

11   *Before the* Quarterly Journal   D. Anderson, Alcohol and public opinion, *Quarterly Journal of Studies on Alcohol* 3 (1942): 376–392.

12   *"What are the ideas"*   Ibid., pp. 377–378.

13   *When he went to work*   P. B. Page, E. M. Jellinek and the evolution of alcohol studies: a critical essay, *Addiction* 92(12) (1997): 1619–1637. For a discussion of Jellinek's faked credentials, see R. Roizen, Jellinek's phantom doctorate, *Ranes Report* (n.d.), www.roizen.com/ron/rr11.htm, retrieved December 6, 2007.

13   *Six years after Prohibition*   J. S. Blocker Jr., *American Temperance Movements: Cycles of Reform* (Boston: Twayne, 1989), pp. 130–161.

13   *Puritans drank at church celebrations*   H. G. Levine, The discovery of addiction: changing conceptions of habitual drunkenness in America, *Journal of Studies on Alcohol* 39(1) (1978): 144–151.

13   *"the good creature of God"*   Ibid., p. 145.

13   *"When a drunkard has his"*   J. Edwards, Freedom of the Will, in O. E. Winslow, ed., *Jonathan Edwards: Basic Writings* (New York: New American Library, 1754/1966) p. 203.

14   *"habitual drunkenness should be"*   Levine, p. 152.

14   *Like many patriots*   And, of course, of ours: George Bush, like many on the Christian right, believes that human rights are not an invention but an endowment, and that it is up to Americans to spread the good news that liberty is the gift of God to mankind.

15   *By 1939, the Cleveland branch*   *Alcoholics Anonymous,* also known as the *Big Book,* is now in its fourth edition. It includes an early history of AA. For the connection to the Oxford Group, see Blocker, pp. 139–140.

16   *It was all spelled out*   W. D. Silkworth, Alcohol as a manifestation of allergy, *Medical Record* (1937), http://silkworth.net/silkworth/allergy.html, retrieved December 6, 2007.

16   *"cannot use liquor at all"*   Ibid., p. 4.

16 *"a law of nature"* Ibid.

16 *"accept the situation"* Ibid.

16 *By taking alcoholism entirely* For more on the relationship between the Research Council and the liquor industry, see Blocker, pp. 149–150, and Page, p. 1622.

17 *"FIRST, that alcoholism"* Blocker, p. 148.

18 *But the proponents of the* M. W. Robinson and W. L. Voegtlin, Investigations of an allergic factor in alcohol addiction, *Quarterly Journal of Studies on Alcohol* 13 (1952): 196–200. For a summary of the findings in the 1950s, see E. M. Jellinek, *The Disease Concept of Alcoholism* (Piscataway, NJ: Alcohol Research Documentation, 1960), pp. 33–82.

18 *"The fact that [doctors] are not"* Jellinek, p. 12.

19 *By 1973, all fifty states* For the history of the institutional adoption of the disease model, see Blocker, pp. 149–152.

22 *"The fracture of a"* P. Sedgwick, Illness—mental and otherwise, *Hastings Center Reports* 1(3) (1973): 30.

1. ADDICTION: VISIONS OF HEALING

26 *It's also because after we arrive* For more information on the ritual use of iboga in West Africa, see J. Fernandez, *Bwiti: An Ethnography of the Religious Imagination in Africa* (Princeton, NJ: Princeton University Press, 1982). For the psychedelic adventurer's view of Bwiti initiation rites, see N. Saunders, *In Search of the Ultimate High* (New York: Random House, 2000).

29 *After tricking the czar* K. Fraser, Local potheads shout down Bush's drug czar, *Vancouver Province,* November 21, 2002, p. A6.

31 *The first reported use* J. M. Hoberman and C. E. Yesalis, The history of synthetic testosterone, *Scientific American* 272(2) (1995): 76–81.

32 *"The temptations of the flesh"* M. Weber, *The Protestant Ethic and the Spirit of Capitalism* (London: Routledge, 1992), p. 157.

32 *The fiction is noble in this respect* See C. Taylor, *Sources of the Self* (Cambridge, MA: Harvard University Press, 1989), esp. pp. 456–496.

34 *Lotsof has his own drug company* U.S. Patent nos. 4499096, 4587243, 4587523, 5026697, and 5152994, available at www.freepatentsonline.com, retrieved November 20, 2007.

35 *Add that to the fact* The most recent figure, compiled from industry reports, is $802 million. See J. A. DiMasi, R. W. Hansen,

and H. G. Grabowski, The price of innovation: new estimates of drug development costs, *Journal of Health Economics* 22 (2003): 153–185.

35 *He helped to start*   See P. DiRienzo, D. Beal et al., *The Ibogaine Story: Report on the Staten Island Project,* available at www.cures-not-wars.org/ibogaine/iboga.html, retrieved November 20, 2007.

36 *Taub, who was once a jeweler*   For more information, see www.ibeginagain.org, retrieved November 20, 2007.

36 *You can download*   The *Manual for Ibogaine Therapy* is available at www.ibogaine.org/ibogaine.pdf, retrieved November 20, 2007.

40 *Deborah Mash has the degrees*   For more information, see www .brainbank.med.miami.edu, retrieved February 3, 2008.

43 *In the 1950s*   For the history of the CIA and LSD, see J. Stevens, *Storming Heaven: LSD and the American Dream* (London: Heinmann, 1988), pp. 74–99. For the uses of LSD to treat alcoholism, see H. A. Abramson, ed., *The Uses of LSD in Psychotherapy and Alcoholism* (Indianapolis: Bobbs-Merrill, 1967).

46 *Moob got the tests*   The manual suggests an EKG and a medical history. It refers readers to various resources, including the *Merck Manual,* for interpreting the results.

47 *She has isolated a metabolite*   The chemistry: M. H. Baumann et al., Noribogaine (12-hydroxyibogamine): a biologically active metabolite of the antiaddictive drug ibogaine, *Annals of the New York Academy of Sciences* 914 (2000): 354–368. The effects: D. C. Mash et al., Ibogaine: complex pharmacokinetics, concerns for safety and preliminary efficacy measures, *Annals of the New York Academy of Sciences* 914 (2000): 394–401.

48 *By binding to the nicotinic receptors*   S. Glick et al., 18-Methoxycoronaridine (18-MD) and ibogaine: comparison of anti-addictive efficacy, toxicity, and mechanisms of action, *Annals of the New York Academy of Sciences* 914 (2000): 369–386.

2. Depression: In the Magic Factory

52 *I told him that*   American Psychiatric Association, *Diagnostic and Statistical Manual of Mental Disorders,* 4th ed., text revision (Washington, DC: American Psychiatric Association, 2000), pp. 775–777. Minor Depression is a provisional diagnosis, listed at the back of the *DSM-IV-TR,* where it awaits further studies like

this one. The nine symptoms of Major Depressive Episode are on p. 356 of *DSM-IV-TR*. Running a study on this diagnosis thus has a twofold aim: to provide another FDA-approved *indication* for the drug, which is another avenue of income for the drug companies, and to give Minor Depression medicine's most lucrative imprimatur: the five-digit code that allows doctors to bill insurance companies for treatment.

53 *I'm a quick shopper* For a searchable database of clinical trials, including this one, see www.clinicaltrials.gov, retrieved November 13, 2007.

53 *To a psychiatrist already convinced* J. R. Hibbeln, Fish consumption and major depression, *Lancet* 351 (1998): 1213; G. Parker, et al., Omega-3 fatty acids and mood disorders, *American Journal of Psychiatry* 163(6) (2006): 969–978.

55 *The HAM-D was invented* M. Hamilton, A rating scale for depression, *Journal of Neurology, Neurosurgery and Psychiatry* 23 (February 1960): 56–62. The HAM-D is no longer under copyright protection and is widely available on the Internet, including at http://healthnet.umassmed.edu/mhealth/HAMD.pdf, retrieved November 13, 2007.

56 *So psychiatrists have developed* For more information, see www.scid4.org, retrieved November 13, 2007.

56 *The* DSM-IV's *fifty-one possible* DSM-IV-TR, pp. 345–429.

58 *A colleague of Price's* G. Greenberg, Is it Prozac or placebo? *Mother Jones* (November–December 2003), pp. 76–80.

58 *"novel prescription pharmaceutical product"* For more information, see www.gwpharm.com/sativex.asp, retrieved June 20, 2005.

58 *"side effect"* For more information, see www.gwpharm.com/research_intoxication.asp, retrieved June 20, 2005.

58 *"a load of bollocks"* G. Greenberg, Respectable reefer, *Mother Jones* (November–December 2005), p. 88.

59 *And the FDA has put Sativex* For more information, see www.gwpharm.com/states.asp, retrieved December 8, 2007.

59 *The Quick Inventory of Depressive Symptomotology* Available at www.ids-qids.org/translations/english/QIDS-SREnglish2page.pdf, retrieved November 13, 2007. The way that researchers decide whether these tests can accurately indicate depression is by correlating responses on them to responses on tests already known

to measure depression—a good idea unless there is no anchor at
the end of the chain, in which case you may well simply have created a self-validating semiotic monster.

59   *The Q-LES-Q*   Available at www.edc.pitt.edu/stard/public/
docs/AssessmentForms/QLESQ-Short%20Form-abridged.doc,
retrieved November 13, 2007.

60   *And on the Ryff Well Being Scale*   Available from the author,
Carol Ryff, cryff@wisc.edu. These are Items 3 and 40.

60   *"life is empty"*   QIDS, Item 12.

60   *"for me to voice my own"*   Ryff, Item 32.

60   *"My thinking is slowed down"*   QIDS, Item 13.

61   *They've broken the code*   For more information, see www.cirp
.org/library/ethics/nuremberg/, retrieved December 8, 2007.

63   *In fact, in more than half*   I. Kirsch et al., The emperor's new
drugs: an analysis of antidepressant medication data submitted
to the FDA, *Prevention & Treatment* 5(1) (2002): 13–14. The average advantage of antidepressants over placebos in those trials was
two points on the HAM-D, a result that could be achieved if the
patient ate and slept better. The average improvement in clinical trials is just over ten points, which means, according to Irving
Kirsch, a University of Connecticut psychologist, that nearly
80 percent of the drug effect is actually a placebo effect.

63   *In addition, when it came time*   P. Leber, Approvable action on
Forrest Laboratories, Inc., NDA 20–822 Celexa (citalopram HBr)
for the management of depression, memorandum, Department
of Health and Human Services, Public Health Service, Food and
Drug Administration, Center for Drug Evaluation and Research
(May 4, 1998).

64   *Until there's money to be made*   See A. Leuchter et al., Changes
in brain function of depressed subjects during treatment with placebo, *American Journal of Psychiatry* 159 (2002): 122–129. While
observing the EEGs of patients in an antidepressant vs. placebo
trial, Leuchter and colleagues stumbled on a pattern of brain
activity common to placebo responders. Drug companies were
very interested in this discovery—not because it allowed them to
study the placebo effect, but because it might allow them to identify placebo responders and bounce them out of a trial before it
started.

65   *Depression, the new theory went*   First advanced by D. W. Woolley
      and E. Shaw, A biochemical and pharmacological suggestion
      about certain mental disorders, *Science* 119 (3096) (1954): 587–588.

65   *In 1961, for example*   F. Ayd, *Recognizing the Depressed Patient:
      With Essentials of Management and Treatment* (New York: Grune &
      Stratton, 1961).

66   *Nor have they explained*   D. Healy, *The Antidepressant Era*
      (Cambridge, MA: Harvard University Press, 1997).

66   *In the face of these dismal results*   A good summary of recent
      research can be found in P. Kramer, *Against Depression* (New
      York: Viking, 2005).

68   *A government psychiatrist*   C. A. Zarate, A randomized trial of
      an N-methyl-D-aspartate antagonist in treatment-resistant major
      depression, *Archives of General Psychiatry* 63(8) (2006): 856–864.

68   *In the psychiatric underground*   For more information and per-
      sonal accounts, see www.erowid.org, retrieved November 13, 2007.

75   *"But you had a good response"*   All of which raises the question
      of how the doctors know what kind of follow-up to provide,
      whether to give a drug or not. Later, the lead investigator on the
      study, David Mischoulon, told me that they "take their best guess"
      about whether the subject was on a drug or a placebo. The reason
      for not disclosing my experimental condition, he explained, was so
      that doctors wouldn't detect a pattern in response and thus "break
      the blind." He added that I could indeed find out when the study
      is completed—about five years from now, he estimated.

75   *"If one listens patiently"*   S. Freud, *Mourning and Melancholia*,
      J. Strachey, trans. (New York: W. W. Norton, 1959), p. 11.

3. SEXUAL ORIENTATION: GAY SCIENCE

81   *All the major psychotherapy guilds*   See, for instance, the
      American Psychological Association's Resolution on Appropriate
      Therapeutic Responses to Sexual Orientation, www.apa.org/pi/
      sexual.html, retrieved December 6, 2007.

83   *Kertbeny published a pamphlet*   J. C. Feray and M. Herzer,
      Homosexual studies and politics in the 19th century: Karoly Maria
      Kertbeny, G. W. Pepple, trans., *Journal of Homosexuality* 19(1) (1990):
      23–48; M. Herzer, Kertbeny and the nameless love, H. Kennedy,

trans., *Journal of Homosexuality* 12(1) (1986): 1–26; and J. N. Katz, *The Invention of Heterosexuality* (New York: Plume/Penguin, 1995), pp. 51–55.

83   *Another antisodomy-law opponent*   S. LeVay, *Queer Science: The Use and Abuse of Research into Homosexuality* (Cambridge, MA: MIT Press, 1996), pp. 11–16. For the connection between biological essentialism and law, see N. Miller, *Out of the Past: Gay and Lesbian History from 1869 to the Present* (New York: Alyson Books, 2006), pp. 14–16.

84   *Hirschfeld was an outspoken*   LeVay, p. 19. For Hirschfeld generally, see LeVay, chap. 1.

84   *Even Sigmund Freud*   Freud's letter to the mother of an American homosexual can be found at http://wthrockmorton .com/?p=420&akst_action=share-this, retrieved July 21, 2007.

84   *These therapies were largely*   R. Bayer, *Homosexuality and American Psychiatry: The Politics of Diagnosis* (New York: Basic Books, 1981), pp. 88–100.

84   *Gay activists, some of them psychiatrists*   Ibid., pp. 99–115.

85   *The culprit in SOD*   Ibid., pp. 115–138.

87   *In that country*   T. G. M. Sandfort et al., Same-sex sexual behavior and psychiatric disorders, *Archives of General Psychiatry* 58 (2001): 85–91; T. G. M. Sandfort et al., Same-sex sexuality and quality of life: findings from the Netherlands Mental Health Survey and Incidence Study, *Archives of Sexual Behavior* 32 (2003): 15–22.

87   *Byrd also describes the studies*   J. M. Bailey and N. G. Martin, A twin registry study of sexual orientation, paper presented at the twenty-first annual meeting of the International Academy of Sex Research, Provincetown, Massachusetts (1995). See also J. M. Bailey and R. C. Pillard, A genetic study of male sexual orientation, *Archives of General Psychiatry* 48 (1995): 1089–1096.

87   *Like everyone else here*   R. L. Spitzer, Can some gay men and lesbians change their sexual orientation? 200 participants reporting a change from homosexual to heterosexual orientation, *Archives of Sexual Behavior* 32(5) (2003): 403–415.

87   *The study was full of caveats*   The peer commentary can be read in *Archives of Sexual Behavior,* volume 32, the same issue in which Spitzer's article appeared.

92 *Two rumors crackle the air* G. Schoenewolf, Gay rights and political correctness: a brief history (2006). Originally published at www.narth.org. Although it has been removed from the NARTH Web site, the paper is available at www.splcenter.org/images/dynamic/intel/SchoenwolfEssay.pdf, retrieved July 21, 2007.

93 *After all, he says* For a comprehensive review of the research, see B. Mustanski, M. Chivers, and J. M. Bailey, A critical review of recent biological research on human sexual orientation, *Annual Review of Sex Research* 13 (2002): 89–140. Also see LeVay, chap. 6; and D. Hamer and P. Copeland, *The Science of Desire: The Search for the Gay Gene and the Biology of Behavior* (New York: Simon & Schuster, 1994), chaps. 7–8. For the "gay rams," see H. Phillips, Homosexuality is biological, suggests gay sheep study (2002), www.newscientist.com/article.ns?id=dn3008, retrieved November 17, 2007.

94 *Gay rights lawyers* For the history of the jurisprudence, see J. A. Williams, Re-orienting the sex discrimination argument for gay rights after *Lawrence v. Texas, Columbia Journal of Gender and Law* 14 (2005): 131–163. For the legislative history, see G. Mucciaroni and M. L. Killian, Immutability, science, and legislative debate over gay, lesbian, and bisexual rights, *Journal of Homosexuality* 47(1) (2004): 53–77.

94 *Michael Bailey, a Northwestern University* A. S. Greenberg and J. M. Bailey, Do biological explanations of homosexuality have moral, legal, or policy implications? *Journal of Sex Research* 30(3) (1993): 245–251.

96 *From her data, Diamond* See L. Diamond, A new view of lesbian subtypes: stable versus fluid identity trajectories over an 8-year period, *Psychology of Women Quarterly* 29 (2005): 119–128. Diamond provided me with updated figures not yet published.

97 *"If I had been born"* Bem's wife, Sandra Lipsitz Bem, also a psychologist, has written about their marriage in *An Unconventional Marriage* (New Haven, CT: Yale University Press, 2001).

98 *They know about Daryl Bem* D. J. Bem, Exotic becomes erotic: a developmental theory of sexual orientation, *Psychological Review* 103(2) (1996): 320–335.

99 *In 1997, I wrote a paper* G. Greenberg, Right answers, wrong reasons: revisiting the deletion of homosexuality from the *DSM, Review of General Psychology* 1(3) (1997): 256–270.

4. SCHIZOPHRENIA: IN THE KINGDOM OF THE UNABOMBER

107 *"As I walked away"* Quoted in S. Johnson, Forensic evaluation of Theodore John Kaczynski, January 16, 1998, p. 11. The
evaluation can be found at various Web addresses, including
http://paulcooijmans.lunarpages.com/psy/unabombreport.html,
retrieved November 16, 2007.

107 *It turns out that* E. Hooker, The adjustment of the male overt
homosexual, *Journal of Projective Techniques* 21 (1957): 17–31. See
also R. G. Evans, Sixteen Personality Factor Questionnaire scores
of homosexual men, *Journal of Consulting and Clinical Psychology* 34
(1970): 212–215; and J. H. Gagnon and W. Simon, *Sexual Conduct:
The Social Origins of Human Sexuality* (Chicago: Aldine, 1973).

108 *As a freshman at Harvard* Letter from Kaczynski to author, July
3, 1998.

109 *I used to have* Ibid.

109 *"The principle that risk"* Letter from Kaczynski to Michael
Mello, October 12, 1998.

110 *"The cabin," he said, "symbolizes"* W. Glaberson, Cabin fever:
Walden was never like this, *New York Times,* December 7, 1997,
section 4, p. 5.

111 *He did manage to have some fun* Declaration of David V.
Foster, November 17, 1997, para. 4, http://web.archive.org/web/
20001031195239http://www.unabombertrial.com/documents/
dvfoster111797.html, retrieved February 16, 2008. Declaration
of Karen Bronk Froming, November 11, 1997, para. 13, http://
web.archive.org/web/19990128002713/www.unabombertrial
.com/documents/froming111797.html, retrieved February 16,
2008. R. E. Gur and R. C. Gur, Summary of neuropsychiatric
evaluation of Theodore J. Kaczynski, November 15, 1997. Copy
provided to me by Kaczynski, available upon request.

111 *"Early on in our sessions"* Foster, November 17, 1997, para. 4
According to Gur and Gur's report, Kaczynski repeated these
comments to them three days later.

111 *I was simply laying* Letter from Kaczynski to author, February 7,
1999.

111 *"his paranoia about psychiatrists"* Foster, November 17, 1997,
para. 4.

112 *"symptom-based failure to cooperate"* Declaration of David
Vernon Foster, November 12, 1997, para. 14, http://web.archive

.org/web/19990128181422/www.unabombertrial.com/documents/
psyche_exhibitA.html, retrieved February 16, 2008.

112   *She interviewed Kaczynski*   Johnson, p. 3.

112   *His hardscrabble, Third World life*   Foster, November 12, 1997,
paras. 11–12.

112   *And his failure to accept*   Declaration of Xavier Amador, November
16, 1997, para. 18, http://web.archive.org/web/19990222133716/
www.unabombertrial.com/documents/amador111697.html,
retrieved February 16, 2008.

113   *You also have to have*   DSM-IV-TR, pp. 312–313.

113   *"In Mr. Kaczynski's case"*   Johnson, p. 41.

114   *Even worse, "He does not challenge"*   Johnson, pp. 27–28.

115   *In* Industrial Society and Its Future   The *Unabomber Manifesto*
is widely available online, including at cyber.eserver.org/unabom
.txt. Retrieved April 25, 2008.

115   *"mind-forg'd manacles"*   From "London," in D. V. Erdman,
ed., *The Complete Poetry and Prose of William Blake* (Berkeley:
University of California, 1982), p. 26.

116   *"The only way out"*   Unabomber Manifesto, para. 140.

116   *"It would be better to dump"*   Unabomber Manifesto, para. 179.

117   *"There's a little bit"*   R. Wright, The evolution of despair, *Time*
(August 28, 1995), p. 61.

117   *"The shift in public image"*   W. Glaberson, Rethinking a myth:
"Who was that masked man?" *New York Times,* January 18, 1998,
section 4, p. 6.

122   *As it happened, I had*   M. Mello, *The United States of America versus Theodore John Kaczynski: Ethics, Power and the Invention of the
Unabomber* (New York: Context Books, 1999).

122   *Yesterday evening I read*   Letter from Kaczynski to author,
November 5, 1998.

124   *It's Exhibit 9*   For more information, see http://web.archive.org/
web/20000614070446/http://www.contextbooks.com/TJK2255/
TJK2255.html, retrieved May 25, 2008.

5. BRAIN DEATH: AS GOOD AS DEAD

130   *Five minutes later, Barnard opened*   The ultimate operation, *Time*
(December 15, 1967). See also M. A. Devita et al., History of organ
donation by patients with cardiac death, in R. M. Arnold et al.,

eds., *Procuring Organs for Transplant* (Baltimore: Johns Hopkins University Press, 1995), p. 20.

131 *But, as he later explained* C. Barnard, Reflections on the first heart transplant, *South African Medical Journal* 72 (1987): 19–20.

131 *"I did not want to touch"* W. Colby, *Unplugged: Reclaiming the Right to Die in America* (New York: AMACOM, 2007), pp. 65–66.

131 *"the child sneezed"* 2 Kings 4:35 (New King James Version).

132 *Over the next three centuries* See A. B. Baker, Artificial respiration: the history of an idea, *Medical History* 15(4) (1971): 336–351. Also, J. L. Price, The evolution of breathing machines, *Medical History* 16(1) (1962): 67–72.

132 *The advent of electricity* J. Gorham, A medical triumph: the iron lung, *Respiratory Therapy* 9(1) (1971): 71–73.

133 *Powered by a quarter-horsepower* P. Drinker and L. A. Shaw, An apparatus for the prolonged administration of artificial respiration: I. A design for adults and children, *Journal of Clinical Investigation* 7(2) (1929): 229–247.

133 *It had glass viewports* A good photo of Drinker's earliest iron lung can be found at http://historical.hsl.virginia.edu/ironlung/pg4.cfm, retrieved March 13, 2008.

133 *Drinker figured out* Drinker and Shaw, An apparatus, part I, pp. 235–236.

133 *Drinker paralyzed cats* L. A. Shaw and P. Drinker, An apparatus for the prolonged administration of artificial respiration: II. A design for small children and infants with an appliance for the administration of oxygen and carbon dioxide, *Journal of Clinical Investigation* 8(1) (1929): 33–46.

133 *He got Consolidated Gas's rescue crew* Drinker and Shaw, An apparatus, part II, pp. 237–243.

133 *Eventually, Drinker even described* P. Drinker and E. L. Roy, The construction of an emergency respirator for use in treating respiratory failure in infantile paralysis, *Journal of Pediatrics* 13(1) (1938): 71–74.

133 *But the first iron lung patient* P. Drinker and C. F. McKhann, The use of a new apparatus for the prolonged administration of artificial respiration: I. A fatal case of poliomyelitis, *JAMA* 92(20) (1929): 1658–1660.

134 *They were, the doctors said* P. Mollaret and M. Gouland, Le coma dépassé, *Revue Neurologique* 101(1) (1959): 3–15.

134 *"patients stacked up waiting for"* D. Rothman, *Strangers at the Bedside* (Piscataway, NJ: Aldine Transaction, 2003), p. 160.

134 *"Can society afford"* *Proceedings of the Conference on the Ethical Aspects of Experimentation on Human Subjects* (Boston: American Academy of Arts and Sciences, 1967), pp. 50–51. See also H. Jonas, Philosophical reflections on experimenting with human subjects, *Daedalus* 98 (1969): 219–245; and H. K. Beecher, Scarce resources and medical advancement, *Daedalus* 98 (1969): 275–313.

134 *"Responsible medical opinion"* A definition of irreversible coma: report of the ad hoc committee of the Harvard Medical School to examine the definition of brain death, *JAMA* 205(6) (1968): 87.

135 *In the decade following* President's Commission for the Study of Ethical Problems in Medicine and Biomedical and Biobehavioral Research, *Defining Death: Medical, Legal and Ethical Issues in the Determination of Death* (Washington, DC: U.S. Government Printing Office, 1981), p. 61.

135 *Practically, it meant that* Ibid., p. 68.

135 *In 1980, a commission appointed* Ibid., pp. 2–8.

135 *But such quality of life criteria* Ibid., pp. 38–40.

136 *In addition, the president's commission said* Ibid., pp. 33–38.

138 *Indeed, a 2008 Alaska legislative* www.abionline.org/pressReleasedetail.cfm?release=73, retrieved May 23, 2008.

139 *In 2007, it helped to manage* For more information, see www.donors1.org/index.php?option=com_content&task=view&id=232&Itemid=73, retrieved February 16, 2008. For national statistics, see www.unos.org, retrieved February 16, 2008.

141 *"the body was made"* Price, p. 68.

143 *Shewmon's inquiry has led him* For an autobiographical account, see D. A. Shewmon, Recovery from "brain death": a neurologist's apologia, *Linacre Quarterly* (February 1997): 31–96.

148 *Non-heart-beating protocols* R. M. Arnold et al., Back to the future: obtaining organs from non-heart-beating cadavers, in R. M. Arnold et al., eds., *Procuring Organs for Transplant* (Baltimore: Johns Hopkins University Press, 1995), p. 3. See also Committee on Non-Heart-Beating Transplantation, *Non-Heart Beating Organ Transplantation* (Washington, DC: National Academy Press, 2000).

150   *"Death is certain"*   Quoted in J. Bondesen, *Buried Alive: The Terrifying History of Our Most Primal Fear* (New York: W. W. Norton, 2001), p. 53.

151   *Doctors got involved*   M. S. Pernick, Back from the grave: recurring controversies over defining and diagnosing death in history, in R. M. Zaner, ed., *Death: Beyond Whole-Brain Criteria* (Dordrecht, Netherlands: Kluwer Academic, 1988), pp. 17–21. Also G. K. Behlmer, Grave doubts: Victorian medicine, moral panic, and the signs of death, *Journal of British Studies* 42 (2003): 210–211.

151   *But it took on urgency*   J.-J. Bruhier, *The Uncertainty of the Signs of Death, and the Danger of Precipitate Interments and Dissections,* 2nd ed. (London: Longmans, 1752). Also see Bondesen, pp. 51–71.

151   *Legend has it that*   S. Nuland, *Doctors: The Biography of Medicine* (New York: Alfred A. Knopf, 1988), p. 92.

152   *"Why, you are in the dead cart"*   D. Defoe, *A Journal of the Plague Year* (London: Penguin, 2003), p. 158.

152   *And, according to Jan Bondesen's* Buried Alive   Bondesen, pp. 32–33.

152   *And then there were the safety coffins*   Ibid., pp. 72–137. Also Pernick, pp. 29–37.

152   *In this case, the public feared*   Behlmer, pp. 224–228.

152   *Indeed, while there may*   T. Marshall, *Murdering to Dissect: Grave-Robbing, Frankenstein and the Anatomy Literature* (Manchester, UK: Manchester University Press, 1995).

153   *When a doctor determines brain death*   Practice parameters: determining brain death in adults, *Neurology* 45(5) (1995): 1012–1014.

153   *Gail Van Norman, who teaches*   G. Van Norman, A matter of life and death: what every anesthesiologist should know about the medical, legal, and ethical aspects of declaring brain death, *Anesthesiology* 91 (1999): 275–287.

155   *During a break between sessions*   See J. Lizza, *Persons, Humanity, and the Definition of Death* (Baltimore: Johns Hopkins University Press, 2006).

6. PERSISTENT VEGETATIVE STATE: BACK FROM THE DEAD

161   *The doctors, in turn*   For a review of these, see J. L. Bernat, Chronic disorders of consciousness, *Lancet* 367(9517) (2006): 1181–1192. Also J. T. Giacino, The minimally conscious state: defining

the borders of consciousness, *Progress in Brain Research* 150 (2005): 381–395.

162   *"Empirical reductionism is in essence"*   C. R. Woese, A new biology for a new century, *Microbiology and Molecular Biology Reviews* 68(2) (2004): 173–186.

164   *He's seen people improve*   J. B. Cooper et al., Right median nerve electrical stimulation to hasten awakening from coma, *Brain Injury* 13(4) (1999): 261–267. See also E. B. Cooper et al., Electrical treatment of reduced consciousness: experience with coma and Alzheimer's disease, *Neuropsychiatric Rehabilitation* 15(3/4) (2005): 389–405.

167   *Doctors at other hospitals*   T. Yamamoto et al., Deep brain stimulation therapy for a persistent vegetative state, *Acta Neurochirugia Wien* 79 (2001): 79–82.

169   *Devastating brain injuries*   K. J. Becker et al., Withdrawal of support in intracerebral hemorrhage may lead to self-fulfilling prophecies, *Neurology* 56 (March 2001): 766–772.

170   *They have tested various hypotheses*   N. D. Schiff et al., Behavioural improvements with thalamic stimulation after severe traumatic brain injury, *Nature* 448(2) (2007): 600–603.

171   *"should motivate research to"*   Ibid., p. 603.

7. MORTALITY: WE'LL ALL WAKE UP TOGETHER

175   *The geoduck's (pronounced "gooey-duck") siphon*   Judge for yourself at www.wdfw.wa.gov/fish/shelfish/beachreg/2clam.htm, retrieved November 15, 2007.

182   *The mind reels at the prospect*   D. DeLillo, *White Noise* (New York: Penguin Books, 1985).

182   *"A flourishing human life"*   President's Council on Bioethics, *Beyond Therapy: Biotechnology and the Pursuit of Happiness* (Washington, DC: Dana Press, 2003), p. 299. Also available at www.bioethics.gov, retrieved March 2, 2008.

184   *Kass's positions on these matters*   See, for instance, B. Joy, Why the future doesn't need us, *Wired* (April 2000), 89–93.

184   *"insights mysteriously received"*   L. R. Kass, *Life, Liberty, and the Defense of Dignity* (San Francisco: Encounter Books, 2002), p. 183.

184 *"We should pay attention"* L. R. Kass, *The Beginning of Wisdom: Reading Genesis* (New York: Free Press, 2003), p. 243.

184 *"the truths he traces"* L. Vogel, Natural law Judaism? The genesis of bioethics in Hans Jonas, Leo Strauss, and Leon Kass, *Hastings Center Report* 36(3) (2006): 32–44.

185 *In Strauss's world* See L. Strauss, *The Rebirth of Classical Political Rationalism,* T. L. Pangle, ed. (Chicago: University of Chicago Press, 1989). See also L. Strauss, Why we remain Jews: can Jewish faith and history still speak to us?" in *Leo Strauss: Political Philosopher and Jewish Thinker,* Kenneth L. Deutsch and Walter Nicgorski, eds. (Lanham, MD: Rowman & Littlefield, 1994), pp. 60–73.

185 *"Having heard it remarked"* The letter was originally written in French. A translated version can be found in A. H. Smith, ed., *The Writings of Benjamin Franklin,* vol. 6 (New York: MacMillan, 1906), p. 42. See also J. J. Gold, Dinner at Doctor Franklin's, *Modern Philology* 75(4) (1978): 391–393.

186 *"I wish it were possible"* Ibid.

186 *Cold, Power concluded did not* See M. Sittig, *Cryogenics: Research and Applications* (New York: D. Van Nostrand, 1963), p. 27.

186 *Indeed, early thermal explorers* See A. S. Parkes, *Marshall's Physiology of Reproduction,* 3rd ed. (London: Longman's, 1960), p. 37.

186 *In 1848, William Thompson* See H. Chang, *Inventing Temperature: Measurement and Scientific Progress* (Oxford, UK: Oxford University Press, 2004).

187 *"One and the same organism"* B. J. Luyet and P. M. Gehenio, *Life and Death at Low Temperatures* (Normandy, MO: Biodynamica, 1940), p. 11.

187 *"like a watch that has unwound"* Deep freeze, *Time* (April 28, 1952), www.time.com/time/magazine/article/0,9171,816378,00.html, retrieved April 29, 2008.

187 *While Luyet's solution* Ibid.

188 *Further experiments proved* C. Polge et al., Revival of spermatozoa after vitrification and dehydration at low temperatures, *Nature* 169 (1949): 666.

188 *In* The Prospect of Immortality Robert C. W. Ettinger, *The Prospect of Immortality* (Garden City, NY: Doubleday, 1964).

188   *In 1965, Wilma Jean McLaughlin's husband*   M. Perry, The first suspension, *Cryonics* (July 1991), available at www.alcor.org/Library/html/BedfordSuspension.html, retrieved April 11, 2008.

188   *Dandridge Cole, a scientist*   Ibid.

188   *"In the end the family did"*   Quoted in ibid.

189   *"There is little or no thought"*   From *Freeze-Wait-Reanimate* (May 1966), quoted in ibid.

189   *There he was placed*   For the complete story of Bedford's cryopreservation, see R. F. Nelson, *We Froze the First Man* (New York: Dell, 1968).

190   *But at least he escaped*   Fully recounted in M. Perry, Dear Dr. Bedford (and those who will care for you after I do): an open letter to the first frozen man, *Cryonics* (July 1991), available at www.alcor.org/Library/html/BedfordLetter.htm, retrieved November 15, 2007.

191   *"nursemaids's lullaby"*   S. Freud, *Civilization and Its Discontents,* J. Strachey, trans. and ed. (New York: W. W. Norton), p. 39.

192   *It is, after all, a variant*   B. Pascal, *Pensees,* A. Krailsheimer, trans. and ed. (London: Penguin, 1995), pp. 121–122.

192   *"The spiritual status of cryonics"*   For more information, see www.alcor.org, retrieved October 15, 2007.

193   *We've heard from the scientists*   If you want to try cryopreservation at home, the recipe is posted at www.nanotechnology.com/blogs/steveedwards/2005_12_01_archive.html, retrieved November 12, 2007. For a scientific look at M-22 by its inventors, see G. M. Fahy et al., Cryopreservation of complex systems: the missing link in the regenerative medicine supply chain, *Rejuvenation Research* 9(2) (2006): 279–291.

195   *He's trying to explain his program*   For more information, see www.sens.org, retrieved November 14, 2007.

195   *He's raised more than $8 million*   For more information, see www.mprize.org, retrieved November 14, 2007.

196   *With each replication*   See S. S. Hall, *Merchants of Immortality: Chasing the Dream of Human Life Extension* (Boston: Houghton Mifflin, 2003), pp. 14–41.

197   *They shake their heads*   The case is *Thomas Donaldson et al., Plaintiffs and Appellants v. Daniel E. Lungren as Attorney General,*

etc., et al., Defendants and Respondents, available at http://online
.ceb.com/calcases/CA4/2CA4t1614.htm, retrieved August 12, 2007.

197 *Of course, Avianna may be* S. Vyff, *21st Century Kids*
(Huntersville, NC: Warren Publishing, 2007).

200 *You will have suffered*   See R. Merkle, Molecular repair of the
brain, available at www.alcor.org/Library/html/MolecularRepair
OfTheBrain.htm, retrieved November 14, 2007.

AFTERWORD

207 *In January, the* New England Journal of Medicine   E. H.
Turner et al., Selective publication of antidepressant trials and its
influence on apparent efficacy, *New England Journal of Medicine,*
358(3) (2008): 252–260.

207 *A Canadian study*   C. Barbui et al., Effectiveness of paroxetine
in the treatment of acute major depression in adults: a systematic
re-examination of published and unpublished data from randomized trials, *Canadian Medical Association Journal* 178(3) (2008):
296–305.

208 *Also in January*   A. Berenson, Drug approved. Is disease real?
New York Times, January 14, 2008, p. A1.

209 *The analysis showed that*   I. Kirsch et al., Initial severity and antidepressant benefits: a meta-analysis of data submitted to the Food
and Drug Administration, *PLoS Medicine* 5(2) (2008): 260–268.

210 *Another front-page story*   J. McKinley, Surgeon accused of speeding
a death to get organs, *New York Times,* February 27, 2008, p. A1.

212 *"In 10 years"*   J. Interlandi, What addicts need, *Newsweek* (March
3, 2008). Available at www.newsweek.com/id/114716, retrieved
February 28, 2008.

212 *But by far the most spectacular*   The transcript and the video of
the hearing, which was held on February 13, 2008, are available
at http://oversight.house.gov/story.asp?ID=1743, retrieved March
2, 2008.

214 *"human nature is on the operating table"*   L. R. Kass, *Life, Liberty
and the Defense of Dignity: The Challenge for Bioethics* (San
Francisco: Encounter Books, 2002), p. 76.

214 *"Truth hath no confines"*   H. Melville, *Moby-Dick* (Boston:
Houghton Mifflin, 1956), p. 139.

INDEX

hormones, 31

Hughes, Jay, 200–201

human nature, 32, 214

hydrazine, 65

ibogaine

Emery's Iboga Therapy House, 26–28, 29, 36–40, 46–47

Glick's research, 48–49

Lotsof's discovery, 20–21, 32–35, 41

Mash's Healing Visions Institute, 40–46, 47

Taub's treatments, 35

Iboga Therapy House, 26, 36–40, 46–47

identical twin studies, 87

identity, 84, 94, 201–2

immortality

historical pursuits of, 185–90

life extension technology, 194–97

President's Council on Bioethics on, 183–84

See also cryonics

Immortality: Physically, Scientifically, NOW (Cooper), 188

inebriety, 11–20

information-theoretic death, 200–202

Institute of Medicine, 148

insurance companies, 129

iron lung, 132–34

Ivey, Candice, 157–60, 161, 172–73

Ivey, Elaine, 158–59, 172–73

Japan, 165–68, 171

Jellinek, Elwin, 12–13, 17, 18

John Paul II, 146–47

Johnson, Sally, 112–14

Journal of the American Medical Association, 134–35

Journal of the Plague Year (Defoe), 151–52

Kaczynski, Ted

appeal motion, 123–24

author's attempt to interview, 120–23

author's correspondence with, 103–5, 108–9, 114–15, 118–19, 124

early journal entries of, 106–7

mental-defect defense, 109–12

our similarities to, 125–26

schizophrenia diagnosis, 112–14, 117–18

Truth Versus Lies, 119–20

Unabomber Manifesto, 115–17

Kanno, Tetsuo, 167, 171, 172, 173

Karten, Elan, 88

Kass, Leon, 183–85, 214

Kertbeny, Karoly Maria, 83–84, 101–2

ketamine, 68–69

Kirsch, Irving, 209

Klein, Donald, 57

Laennec, René-Théophile-Hyacinthe, 1–2, 3

Leary, Timothy, 32, 44

Life Enjoyment and Satisfaction Questionnaire (Q-LES-Q), 59–60

Life (magazine), 189

life span, 195–97

life support, 131–34, 141–42